RAPID INFORMATION SYSTEMS DEVELOPMENT:

A non-specialist's guide to
analysis and design in an
imperfect world

Simon Bell

Trevor Wood-Harper

McGRAW-HILL BOOK COMPANY

London · New York · St Louis · San Francisco · Auckland · Bogotá
Caracas · Hamburg · Lisbon · Madrid · Mexico · Milan · Montreal
New Delhi · Panama · Paris · San Juan · São Paulo · Singapore
Sydney · Tokyo · Toronto

Published by
McGRAW-HILL Book Company Europe
Shoppenhangers Road, Maidenhead, Berkshire, SL6 2QL, England
Telephone 0628 23432
Fax 0628 770224

British Library Cataloging in Publication Data

Bell, Simon George, 1957-
   Rapid information systems development. (International
software engineering).
   I. Title   II. Wood-Harper, A. T.   III. Series
001.5

   ISBN 0-07-707579-X

Library of Congress Cataloguing-in-Publication Data

Bell, Simon, 1957–
   Rapid information systems development: a non-specialist's guide to analysis and design in an imperfect
world/Simon Bell, Trevor Wood-Harper.
     p. cm. – (The McGraw-Hill international series in software engineering)
     Includes bibliographical references and index.
     ISBN 0-07-707579-X: £21.95
     1. System design. 2. System analysis. I. Wood-Harper, A. T.
   II. Title. III. Series.
   QA76.9.S88B44   1992
   004.2'1 – dc20
                                       91-42538
                                          CIP

12345 95432

Typeset by Alden Multimedia Ltd.

and printed and bound in Great Britain by Clays Ltd, St Ives plc

For Ellen, Rachel, Rebecca and
Jonathan, our children.

Recognize the unity in multiplicity.

CONTENTS

FOREWORD

This book is intended as a practitioner's guide to those non-experts who are intending to plan and develop information systems, i.e. become involved with the process of systems analysis and systems design (SA&SD). The authors recognize that many other approaches are possible in this complex and evolving field, and that greater depth of understanding than that which arises from the reading of one book will be required before exponents could be said to have achieved mastery of all the techniques included here. Nevertheless, the authors believe that at present there is a lack of understanding in the information system planning profession of the need for planning tools for non-computer specialists, and that these tools do exist and can be understood and applied by non-specialists relatively quickly. This book should be seen as an introduction to the information systems development process and as a guide to one particular method. It is to be hoped that this may encourage more professionals working in this field to write training materials of value to non-specialists.

The authors welcome any constructive comments and observations arising from the application of principles contained in this text, especially from users working in situations of rapid change and minimal time for long drawn-out development procedures. This book is written in recognition of the need to draw together 'clean' theory and what is often 'dirty' practice in one view.

The examples used in this book are amalgams brought together from field experience, theory and teaching. They do not represent any one single situation. Any resemblance to any real organization is coincidental.

PREFACE

It is not the authors' intention to produce a work of pure systems analysis and systems design theory. If this is what the reader is hoping to find, then he or she will be disappointed. Nor is it our intention to provide readers with an idealized analysis and design procedure. This book is about doing systems analysis and systems design under conditions where the only alternative to rule-of-thumb methods is to not use any methods at all. This book is intended for those whom the information systems profession would refer to as non-specialist. It is aimed at assisting users in doing the preparatory work (called systems analysis and systems design) that should occur before an information system is installed. We wish to assist those involved in doing this work because, to date, there has been little support for them in their travail unless they proposed undertaking a three-year university degree.

This book does not contain pure examples of applied methodology. Almost all the examples discussed here are drawn from work undertaken in the rigorous environment of developing countries where computer awareness and computer systems development is in its infancy. Therefore the analysis and design tools discussed here have had to be adopted and adapted rapidly when there has been little time and in low support environments (low support in terms of poor climatic conditions, poor infrastructure and low awareness). Nevertheless, we believe that an adapt and adopt approach used in imperfect conditions is better than no approach to planning at all, and it is with this in mind that the following is offered.

The book contains examples of a number of analysis and design approaches, including Multiview itself, which have been adapted from their pure condition to meet the needs of variable, real situations. To the authors of these approaches we offer our apologies but also invite their consideration of our central point: 'To provide an analysis and design tool for non-experts one must occasionally simplify the tool'.

ACKNOWLEDGEMENTS

The authors wish to express their gratitude to I. Antill for inputs to earlier, collaborative work (1985) which provided inspiration for the present volume. Thanks also to Martin Sewell who provided much support in the application of this approach in developing countries, Ian Shephard for his comments and insights, and Professor Richard Briston for his inspiration, and the opportunity to develop Information Systems ideas in developing countries.

Words fail to express the gratitude which the authors owe their families for their support and patience, in particular Ted Bell for his primary edit of the original manuscript and Rachel Furze for her final proof reading.

INTRODUCTION TO THE BOOK STRUCTURE AND CONTENTS

This book attempts to provide the user with an easily understandable set of rules of thumb for planning an information system when there are no 'experts' available to fall back on. As a 'user's guide' this text does not go into great detail concerning the theoretic context of the planning tools we apply. A list of theoretic publications is included in the text.

The book is organized so as to reduce the amount of time the reader has to spend on areas that may not be of immediate interest. Implicit in the approach we are adopting here is the understanding that in many situations where information systems are required there is not always a lot of time to carry out in-depth planning. Because of this a *rapid information system planning or development technique is required* to enable organizations to develop effective information systems.

With this in mind we will briefly outline the structure of the book.

The chapters are organized so as to cover the sequence of activities involved in planning any potential information system. As we are chiefly dealing with the *practical details* of planning, unnecessary *theory* that may be of interest but is not central to this theme is located in *theory appendices* at the back of the book. Each appendix provides the reader with further insights into the subject and lists key texts that can be pursued for further reference.

At the start of each chapter is a listing of the *major keywords* to be dealt with in the body of the chapter. Each chapter opens with a brief *summary* of contents and an indication of how long each stage of the method applied would take to carry out in practice.

The chapters that deal with the fivefold aspects of the method we are going to use are each closed with a *tutorial* or *sequence of exercises*. These are intended to be of value if the book is to be used directly as a stimulus to the planning process in an organization or set as an introductory text in analysis and design teaching.

Our final appendix sets out a model approach to answering the exercises in the tutorials.

The chapters contain an explanation of the working of our analysis and design methodology and contain examples of each stage.

Much of the vocabulary used in information systems and information technology related areas consists of jargon. Some jargon is useful and some is unavoidable. In order to assist the non-specialist, all major jargon and abbreviations used in the text are covered in the *glossary*. Words that appear in the glossary are identified on their first appearance in the text by appearing in bold type.

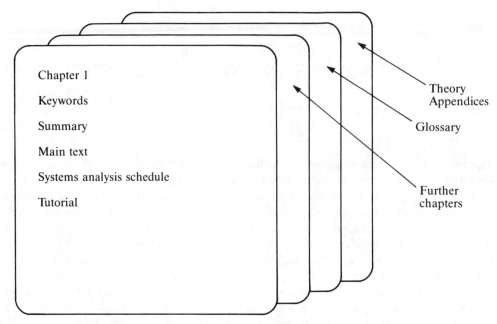

Chapter 1

Keywords

Summary

Main text

Systems analysis schedule

Tutorial

Theory
Appendices

Glossary

Further
chapters

Overall structure of the book.

As a guide, we have provided a schedule of the progress of the planning process as we go through. This is not intended to be absolute but should provide the user with a rough guide to the amount of time to be devoted to each activity in the planning process.

The journey of a thousand miles begins with a single step.
(Lao Tze, 550 BC)

It's not in devising the system,
That the fearful dangers lurk.
It's not in devising the system,
But making the system work.
For the working out of the system,
Is not in the hands of the great,
But rests in the hands of the poor little clerks,
Like Ahmed and José and Kate.
(Anon, 1990)

INFORMATION SYSTEMS AND ORGANIZATIONS

Keywords planning, change, risk, organization, method.

Summary Information as a commodity is briefly discussed. The need for information systems planning is introduced and described. Common problems with information systems are reviewed.

1.1 INTRODUCTION

It may appear to be obvious but **information systems** are supposed to inform people. In the planning or development process we should never lose sight of this primary objective. By informing, the system assists people (generally referred to in the computer context as 'the users') to make intelligent decisions. Therefore, if information is:

- poorly gathered and sorted, inadequately edited, incorrectly analysed, analysed for the wrong things, and badly presented

the information system will fail in its primary function. This in turn has a knock-on effect on decision making, the results of which feed through to the effectiveness of the organization as a whole. Therefore any information system needs to be carefully planned in terms of:

1. The **data** to be gathered.
2. The **information** products being derived from the data.
3. The ultimate **knowledge** that is thought by the planner to be the final requirement of the system (a very difficult thing to define).

All too often an information system is designed prior to anyone having asked the question that it is intended to answer. Thus, incorrect data is gathered, inappropriate information products are generated, and insufficient knowledge is derived for effective decision making.

Organizations of all kinds, be they small private companies or large government departments, are primarily users and producers of information. Information is a most versatile and pliable commodity. Literally anything that leads to any form of action could be seen as being information. A kick in the rear or a bank statement will lead to action, immediate or delayed, positive or negative, brief or sustained. It is worth briefly describing some of the major attributes of information systems:

1. They deal with an endlessly changing commodity.
2. They are required to facilitate decision making.
3. They exist in all organizations.
4. They are vital to an organization's function.
5. They are increasingly available in a computer-based form.

An information system, particularly a **computer-based information system**, can appear to be efficient and yet be perceived by end users as being unhelpful or even hostile. There can be many reasons for this, some of which we look at in Section 1.2. At this point we need to make clear that an information system is an integral part of the social system which comprises another part of the organization (see Fig. 1.1).

The way in which an organization functions is very complex and information systems impinge upon most of the features of an organization. Therefore, as well as being established on technically sound principles, the planning process needs to be both diligent and sensitive to organizational needs and user thinking.

By 'organizational needs and user thinking' we mean such common issues as:

1. Lack of experience of the planning or **systems analysis and systems design** process in the **recipient community**.
2. Reluctance on the part of senior management to adopt suggested change.
3. Reluctance on the part of junior staff to change.
4. Absence of local, reliable support for the incoming system.
5. Sense of risk and uncertainty in a new endeavour.

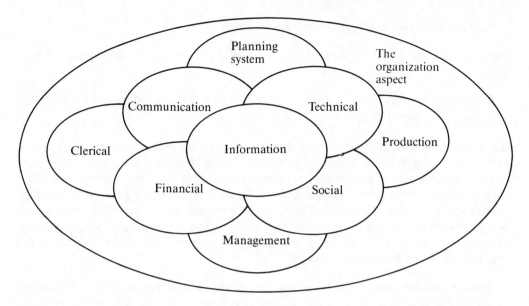

Figure 1.1 Linked aspects.

6. Staff too stretched to accommodate changes gladly.
7. Distrust of all things new.

While not wishing to give the impression that information systems in use today are generally unfriendly and inappropriate, it is our experience that there are a large number of such systems which fail because they do not take into account the views and worries of the end users.

This book is designed to introduce a means for effective information systems planning for organizations, while taking into account this range of issues.

One of the problems that often arises with new computer-based information systems is that users feel that the new information system is being imposed on them with little or no discussion. It is our contention that, generally speaking, the end users of information systems do not have enough say in the analysis and design process. This leads to a lack of sense of ownership among the end users of the system, which is supposedly being implemented to improve user efficiency. Further, the information systems analysis and design process is suffering from a dose of **expert imposition**. To explain what we mean by expert imposition we have to look no further than the types of problem that have been confronting architects in recent years.

After years of quiescence the end user (in this case the home owner or office worker) is asking architects questions like: 'Would *you* live in one of your buildings?' The question being asked of information system planners is:

Would *you* like to work with one of your information systems?

Simply stated, one of the major problems implicit in information systems design appears to be:

- If information systems are planned at all, they tend to be planned by and for computer experts, not general users.

By this we mean that information systems often suffer from a highly technocratic approach derived from experts of the computer profession. Those that use this approach tend to be large companies and/or government departments and agencies which have access to the necessary financial resources required to purchase the professional skills for systems analysis and systems design. This type of planning is characterized by the information system being designed in isolation from the end user in most stages. *Systems for smaller organizations usually develop on a evolutionary, piecemeal basis with little or no overall planning*. The problem can be broken down as follows:

1. Between the computer profession and the general user there is still a considerable knowledge gap. This gap is partly incidental because of the newness of the computer profession, and partly contrived by the computer profession due to a tendency to obscure simple or obvious ideas in confusing jargon.
2. The knowledge gap is a convenient means for information system analysts and designers to keep away the uninitated and the eventual user of the system under design.
3. This tendency leads to professional conceit on the one hand and user mistrust on the other.
4. Yet, returning to our initial problem, users do require working information systems, and they often require them rapidly.

For the majority of organizations without access to professional skills, how are the information systems to be planned? What features of an organization need to be analysed? How is the final information system to be implemented?

If computer experts are not available, are too busy, too intimidating, or cost too much, users tend to fall back on their own means and muddle along. This situation can lead to considerable difficulty and cost, but it is an increasingly apparent tendency. Unplanned or poorly planned systems are on the increase because:

1. There are literally millions of **microcomputer** sites globally that are not run by experts from the computer profession.
2. Contrary to the opinion of some computer experts five to ten years ago, microcomputers, far from being a 'blip', are becoming increasingly sophisticated and undertake an immense range of tasks.
3. Powerful **hardware** and **software** require exceptional and new skills from users.
4. User training has tended to be badly organized and undersubscribed.

5. This has resulted in lack of method in planning information systems and to massively underutilized technology (e.g. powerful microcomputers, intended for sophisticated accounting operations, being used for **word-processing**).

The purpose of this book is to demonstrate an easy-to-use method for identifying what the information system planner (i.e. you) needs to know. Following this, we:

1. Demonstrate a method to use models with the range of technical, economic, social, cultural, political, etc., issues that may be critical to the running of the information system.
2. Produce a model in the definition of the proposed new information system.
3. Identify key technical and social combinations that will achieve the new system products at the right cost.
4. Plan for the interface between users and technology.
5. Outline the major technical processes and facilities that need to be in place for the system to work effectively.
6. Set out software and hardware selection procedures and the implementation process.

We need to make clear now several major features of the book:

- You do not need to be a systems engineer or even know much about computers to be able to make effective use of the book. In fact, this book is intended for non-experts.
- The book deals with information systems design. The end product probably will be automated but *may be semi-automated or even manual*.
- Many people put in charge of new information systems have little previous experience. Therefore the book is aimed at managers who need a simple-to-use, non-technocratic analysis and design tool.
- Existing workers in the profession sometimes doubt the value of existing technical analysis and design tools and sometimes have very limited time to come up with an end product. Therefore, we also intend the book for information systems analysts and designers who need a rapid-use tool.
- Finally, many disciplines make use of information systems but do not always have specialist computer professionals on call to deal with planning. Hence we have made provision for the book to be useful both to a wide range of professionals working in their own disciplines, and lecturers and students interested in bringing some design skills into their specialist area (e.g. management, business, economics, planning).

We hope that all the end users of the book find in it a technically sound but rapidly applicable and socially sensitive planning tool.

1.2 PROBLEMS WITH INFORMATION SYSTEMS

The central theme for this book is the problem of devising a clear and user-sensitive approach to determining exactly what is the problem for which an information system is perceived as being the answer. How do we plan this system and offer a reasonable chance of successful use? With this in mind it is useful to look at some of the standard problems we have encountered in previous systems planning exercises. The types of problem that are in some ways typical of information technology adoption include:

- Patchy understanding of the computer involved by the potential users.
- Very wide range of requirements by the user.
- Awkward environmental factors involved in potential siting of sophisticated systems.
- High cost training and staffing implications.

These points led us to an observation concerning the impact of information systems in organizations where situations of risk and uncertainty prevail:

> Users generally believe that little can be understood about an information system prior to installation. Further, it is often believed that information technology will probably lead to negative rather than positive work experiences.

Taking this as our lead point, our approach is to reassure the user on these counts. But why are users apparently so wary of information systems? Some examples may help to explain.

Our first example in Fig. 1.2 shows that the problem for the analyst is not the data to be

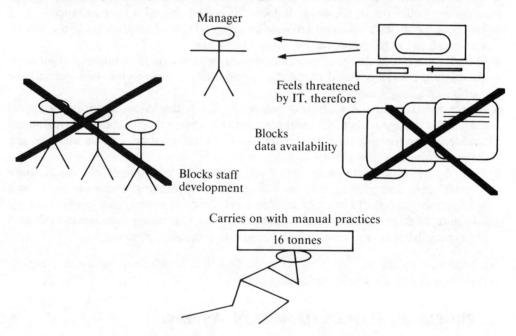

Figure 1.2 Problems with systems. Problem no. 1: senior management intransigence.

prepared or the staff to be trained, but consists in senior management intransigence. If management is left out of the decision-making process and is not included in discussions or consulted, then systems can fail. This demonstrates that it is vital to get the support of major **stakeholders** in the system. In this case the failure results in continuation with existing manual practices which could be easily, technically improved upon.

Lesson 1 would seem to be *always draw management into the analysis and design stage —do not let the technology appear threatening (or any more threatening than it is already!)*.

Figure 1.3 overleaf demonstrates our second problem situation. Here the problem is the planner or system analyst.

In this example the planner devises and suggests a new system but this is quite different from those which the major stakeholders require. Problems arise when an analyst's enthusiasm to create systems that are theoretically sound (in terms of the personal preferences of the analyst) rather than **context**ually appropriate (i.e. what the user wants) predominate. The result is that the analyst is at odds with the preference of the users or clients, and ends up creating conflict and ultimate failure. This situation can be resolved in two ways. Either the client tells the analyst to think again, or, as in this case, the analyst imposes the system. This example is generally restricted to cases where either the planner has been given *carte blanche* to impose his or her will, or if a larger parent or funding body outside the organization has some say in commissioning the study and backs the analyst's judgement.

The result is fairly obvious—a conflict of objectives and ultimately systems failure. The lesson to be learnt is that *irrespective of professionalism the planner/analyst must have the humility and common sense to see the client as central to a working system*.

Our third example is shown in Fig. 1.4 on page 9 and can be seen as a problem of overambition and the price of initial success. Computer systems, like any other system, have to provide their utility to the end user for some considerable time. If they do not provide this utility, they can end up as being more disruptive in their effect than continuing with outmoded and outdated manual practices. In Fig. 1.4 the organization is left with a potentially catastrophic situation where the system ultimately fails but the confidence during the first few years has been so high as to lead to the dismantling of existing manual systems. The lesson is again quite stark: *short-term success can lead to long-term failure unless real long-term support is built into projects. This is a danger for all information system projects.*

Our fourth example demonstrates again the problems of short-term successes. The problem in this case is the overadoption of a computer system. This type of problem can manifest itself in many ways. Examples include microcomputer systems running for over twenty hours a day, seven days a week and printers outputting day and night. In Fig. 1.5 (page 10), the computer-based information project produces such a positive response from the organization that it encourages a massive increase in use.

The lession is obvious: *a system should be designed to meet the needs of today and tomorrow and the next day so far as we can prescribe it!*

Our fifth and final example (Fig. 1.6 on page 11) depicts what might be thought of as the usual problems that a computer-based information system might have to deal with.

The key to almost all these problems is poor security. The requirements for security are

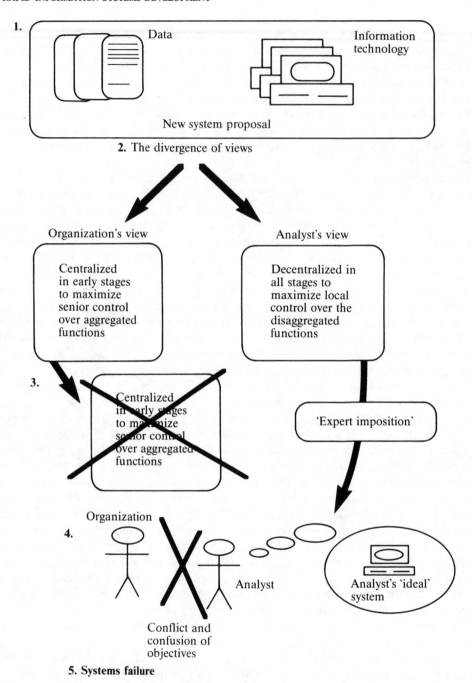

Figure 1.3 Problem no. 2: poor analysis perspective.

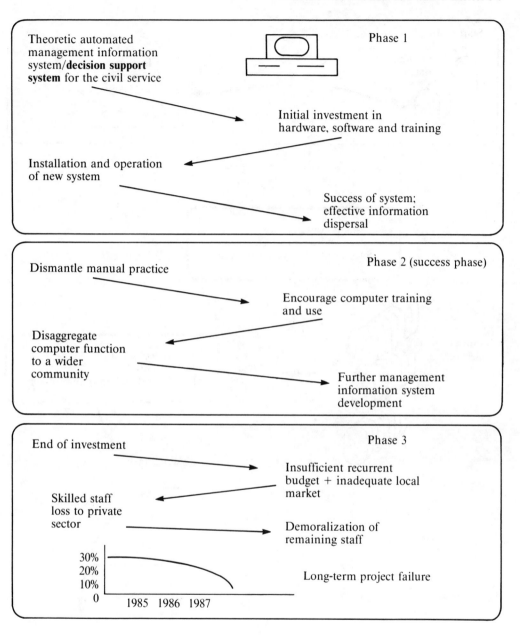

Figure 1.4 Problems with systems. Problem no. 3: overambition.

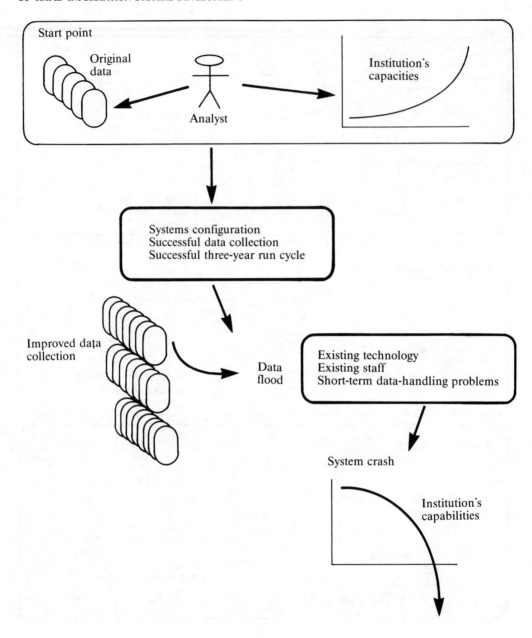

Figure 1.5 Problems with systems. Problem no. 4: task–machine development mismatch.

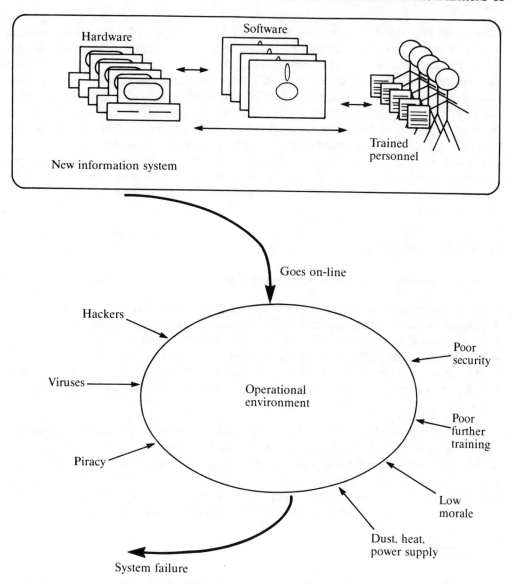

Figure 1.6 Problems with systems. Problem no. 5: technical/management breakdown.

directly related to the specific factors at work in the potential information systems environment. There is generally a trade-off between ease of access to information systems and security. The higher the security, the more difficult the system is to use. Conversely, the more open the system is, the easier it is for computer **hackers** to gain access, for computer

viruses to be imported into the system, and for software to be **pirated**. All of these problems tend to be symptomatic of a less than adequate approach to information systems design. Related issues include insufficient attention being devoted to training of computer staff and the resulting lack of confidence and morale of these staff members. Finally, we must recognize that information systems are often being installed in high-risk environments. Risk varies with situation but can refer to, for example, physical factors such as intermittent power supply; heat or humidity; financial risks such as insufficient recurrent budget for computer support; or social risks such as antipathy to the incoming system.

The lesson is that *it is good practice always to ensure that an information system is going into a hostile environment. Modest systems planned for difficult situations can always be built upon later. Technically sound, ambitious systems may suffer teething problems that take years to recover from.*

1.3 CONCLUSIONS

From the foregoing we can draw some conclusions:

- All information systems exist to support efficient decision making.
- Efficient decision making is vital for personal and organizational well-being.
- Information systems therefore have to be properly planned for.
- Poorly planned or unplanned systems can lead to catastrophe.
- Many planned systems are too technocratic and also lead to problems for the end users.
- A key requirement therefore is for an easy-to-use method for planning information systems.

The usual term used to describe information system planning is 'systems analysis and systems design' or SA&SD. This is rather a mouthful and can appear to be quite an off-putting expression to the non-specialist. However, in order for us to make sense of the planning process we need to understand analysis and design in outline at least. Later on we will be using tools drawn from various types of analysis and design to plan for our own information systems, but first it is useful to know what analysis and design is.

FURTHER READING

Antill, L. and Wood-Harper, A. T. (1985) *Systems Analysis*, Made Simple Computer Books, Heinemann, London.

Argyris, C. (1985) Making knowledge more relevant to practice: maps for action. In E. Lawler *et al.* (eds), *Doing Research that is Useful for Theory and Practice*, Jossey Bass, San Francisco, Calif.

Avison, D. E. and Wood-Harper, A. T. (1991) Information systems development research: an exploration of ideas in practice. *Computer Journal*, **34**, no.2.

Boland, R. (1985) Phenomenology: a preformed approach to research in information systems. In E. Mumford, *et al.* (eds), *Research methods in information systems*, North Holland.

Checkland, P. B. (1985) From optimism to learning: a development of systems thinking for the 1990s. *Journal of the Operational Research Society*, **36**, no.9, pp. 757–767.

Checkland, P. B. (1988) Information systems and systems thinking: timed to unite? *International Journal of Information Management*, no.8, pp. 239–248.

Haynes, M. (1989) A participative application of soft systems methodology: an action research project concerned with formulating an outline design for a learning centre in ICI chemicals and polymers. M.Sc. thesis, University of Lancaster.

Kozar, K. A. (1989) *Humanized Information Systems Analysis and Design: People Building Systems for People*, McGraw-Hill, New York.

Lucas, H. C. (1985) *The Analysis, Design and Implementation of Information Systems*, McGraw-Hill, New York.

Vickers, G. (1981) Systems analysis: a tool subject or judgement demystified. *Policy Sciences*, **14**, pp. 23–29.

Winograd, T. and Flores, F. (1986) *Understanding Computers and Cognition*, Ablex, London.

Wood-Harper, A. T. (1989) Comparison of information systems definition methodologies: an action research, Multiview perspective. Ph.D. thesis, School of Information Systems, University of East Anglia, Norwich.

2

WHAT IS SYSTEMS ANALYSIS AND SYSTEMS DESIGN?

Keywords systems analysis, systems design, research approach, systemic, reductionist, methodologies, tools.

Summary How do we plan an information system with systems analysis and systems design? The experts have produced a vast range of **methodologies** for the planner with a bewildering array of approaches. The incomprehensible language often used and the belief that analysis and design takes months rather than days often invokes the question 'How is any of it of use to us?' In this chapter we look briefly at the range of methodologies and focus in on some major, useful themes.

2.1 INTRODUCTION

Systems analysis and systems design (SA&SD) is the term used to describe the means used to plan an information system. Usually SA&SD is set out within the context of research containing a methodology of some form. Crudely speaking, the methodology will tend to have the following elements.

Basics of a systems analysis and systems design methodology

1. Discover what the information problems are.

2. Discover the setting for the problems.
3. What resources and constraints are evident?
4. What are the major information components of the problems?
5. Structure the problems into a model.
6. Design model solutions to the problems.
7. Test and cost the model.
8. Implement the model as appropriate.

This list of activities looks deceptively simple. An example might be as follows.

Example of a systems analysis and systems design methodology in action

1. Discover what the information problems are. Unacceptable lag between the preparation of departmental budgets and the presentation of these budgets to central financial committee for approval.
2. Discover the setting for the problems. Three major departments are the main offenders —planning, design and maintenance—but all departments are occasionally late with their presentations.
3. What resources and constraints are evident? Central management have indicated a budget of £x on a feasibility study into the problem of procurement of information technology and related staff. There is evidence that young junior staff would be keen to see changes. One negative point is that in the past there has been senior staff intransigence to change and to the perceived whittling down of responsibility and power implicit in a computer-based solution.
4. What are the major information components of the problems? Departmental projects and performance criteria are the main inputs.
5. Structure the problems into a model. This requires the production of an overall plan encompassing an organizational chart of some kind showing key departments in the proposed system and identifying where the existing blockages and delays are with regard to setting projections and performance criteria.
6. Design model solutions to the problems. Once problems in terms of blockages and delays have been identified, model solutions need to be designed which focus initially on the main offenders.
7. Test and cost the model. Depending upon the resources available a thorough bench test of any new model is required. The test would normally take the form of a pilot study with the improved system with key information indicators being monitored for comparison with existing practices, e.g. how long did it take to get manager reports assessed on the computer-based system as opposed to the manual system?
8. Implement the model as appropriate. Implementation will normally follow a successful pilot study and can take a wide variety of forms—e.g. parallel systems, a continued pilot approach or a simple switch over (these and other implementation strategies are described in Chapter 10).

Perceptions
Meaning
Personal view of reality

Technical facts
Objective meaning

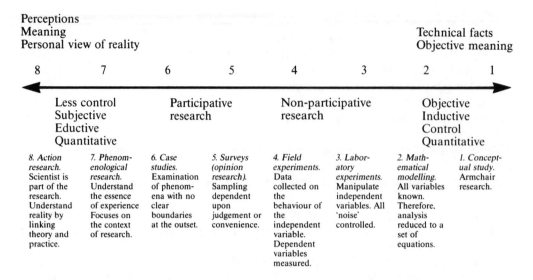

8	7	6	5	4	3	2	1

Less control Subjective Eductive Quantitative		Participative research		Non-participative research		Objective Inductive Control Quantitative	
8. Action research. Scientist is part of the research. Understand reality by linking theory and practice.	*7. Phenom-enological research.* Understand the essence of experience Focuses on the context of research.	*6. Case studies.* Examination of phenom-ena with no clear boundaries at the outset.	*5. Surveys (opinion research).* Sampling dependent upon judgement or convenience.	*4. Field experiments.* Data collected on the behaviour of the independent variable. Dependent variables measured.	*3. Labor-atory experiments.* Manipulate independent variables. All 'noise' controlled.	*2. Math-ematical modelling.* All variables known. Therefore, analysis reduced to a set of equations.	*1. Concept-ual study.* Armchair research.

Figure 2.1 Research methodologies continuum. (Adapted from Wood-Harper, 1989, adapted and extended from Douglas, 1976.)

A vast array of different methods are available for fulfilling this sequence of tasks. The approaches all have their own benefits and weaknesses. Generally they vary from each other along the lines of the research background and training of the individuals who designed them. One way of understanding what is meant by 'research background' can be seen in terms of an axis that shows a range of approaches to research (see Fig. 2.1). Do not panic about this. A few words of explanation may be required! The axis (called a continuum) is shown here with eight points on it. There may be many more points that we could add, but the eight shown here should be sufficient for our example.

To the left are what we refer to as the '*soft*' or generally social sciences based approaches to research; to the right are some of the technocratic, '*hard*' science approaches. All eight of the approaches briefly shown here have salient features making one different from all the others. Each has its own assumptions or **worldview**. This is an important point.

The assumptions of the methods to the right of the continuum are derived from sciences akin to engineering and are focused on a controlled and controllable universe in which science knows all that is needful to know. In contrast, the assumptions of the approaches on the left are based upon the vagaries of human nature, assume that there are very few fixed points upon which the researcher can depend, and often assume that nothing can be absolutely known.

Before we go any further we need to make it clear that understanding and applying research techniques is important for our main task, namely making working information systems. Setting up an information system requires the user to undertake research. To understand problems, deduce the strengths and weaknesses of the environment, plan a new

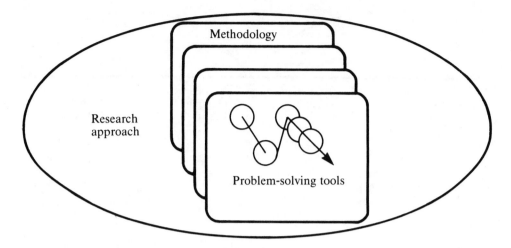

Figure 2.2 Research, methodology, problem-solving tools.

system, and test it prior to implementation requires research skills. The main point that needs to be understood is that most people are quite able to undertake research. All that is required is for the potential researcher to see research in the context of the problem.

For our purposes research refers to the adoption of an overall framework for the application of a methodology and the tools that comprise a methodology. The tools have then to be applied in a sequential and logical manner in order to arrive at an understanding of the problem, some suggestions for improvement, and means for producing the improved situation (see Fig. 2.2).

How does the research–methodology–tools sequence work?

In one example, a management systems planner wants to know if a series of measures aimed at supporting decision making at middle management level will be accepted by staff. Because he is working for a large, multinational company he cannot ask every single member of middle management if they agree with the proposed system. The first priority is therefore to specify a research technique. In this example a case study (point 6 on our axis in Fig. 2.1) approach is used. Certain representative departments are checked. The methodology applied is called participative interaction, which requires the problem-solving tool of questionnaires to be applied.

In another example an agronomist is employed to discover the most appropriate of six seed varieties for the production of maize. She applies the mathematical modelling research technique (point 2 on the axis), uses a sampling methodology, and makes use of tools for measuring leaf growth and seed production among the six varieties.

As planners we need to be sure that we are fully aware of our approach to research, the methodology to be used, and the range of tools which that methodology makes use of. Before we get to this point it is useful first to explain a little more of what the continuum means to us as planners. How do the eight approaches affect us?

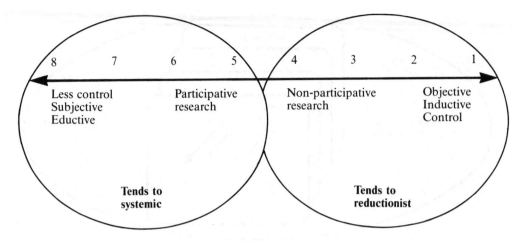

Figure 2.3

The eight-fold division is not definitive and could be added to. However, this range of backgrounds indicates two basic schools of thought or systems of analysis and design. These are known as the hard science school which is often argued to be **reductionist**, and the **systemicists**. Although we go into some greater detail concerning the various methodologies in Appendix 1, it is useful to get an overview of the two camps here (see Fig. 2.3).

2.1.1 The reductionists

Reductionism is the core behind most of the 'hard' sciences and is centred on the philosophical teaching of positivism.

> All genuine human knowledge is contained within the boundaries of science. That is, the systematic study of phenomena and the explication of the laws embodied therein. Philosophy may still perform a useful function in explaining the scope and methods of science, pointing out the more general principles underlying specific scientific findings, and exploring the implications of science for human life. But it must abandon the claim to have any means of attaining knowledge not available to science. (A. Flew, *A Dictionary of Philosophy*, 1988)

With a reductionist approach out go ideas about the reality and importance of 'unscientific' aspects of life (hunches, guesswork, instincts for rightness, and even in certain circumstances illogical activity, i.e. activity that is not consistent with a narrow definition of efficiency). The universe is seen as fixed, knowable, measurable, and, therefore, predictable. Of course this is an ideal definition of positivism but this is the structure of thinking behind three of the analysis and design methods in common use today:

1. Structured systems analysis (SSA).
2. Technical specification (TS).
3. Data analysis (DA).

2.1.2 The systemicists

This is not yet a term to be found readily in dictionaries of philosophy but the approach arises from systems thinking and this will increasingly be related to social philosophy.

> An expression afflicted with the same vagueness and ambiguity as besets 'philosophy' itself, as well as with other troubles of its own. Originally it was applied to any general and comprehensive vision of how society ought to be. . . More recently the same expression has come to be used like 'political philosophy', 'moral philosophy', 'metaphysics', and 'epistemological'. (A. Flew, *A Dictionary of Philosophy*, 1988)

Systemicists are involved in the necessarily subjective world of real human activity. Central to systemisism is the belief that social and political forces will and must interfere with any technocratic information system. The information system planner imposes opinions and beliefs upon logical and objective new systems that are being planned. The systemic view of reality is characterized by an inter- and trans-disciplinarian approach, i.e. linking together various sciences and approaches. Again we have three mainline approaches which we might use to illustrate its working in practice:

1. General systems theory (GST).
2. Soft systems methodology (SSM).
3. Socio-technical systems (STS).

(Refer to Appendix 1 if you want more detail concerning any of these three approaches.)

2.2 WHAT IS OUR RESEARCH APPROACH AND METHODOLOGY?

This is the most subjective question to ask. Implicit in the continuum shown in Fig. 2.1 is the observation that all methodologies have their strengths and weaknesses. It is not a question of selecting the 'right' methodology, rather we believe that a better approach is to select the *right combination of methodological tools for the particular situation* in which you are working.

Before going on to this, however, we should define the research approach of this book. Generally speaking we assume that you, the planner/analyst and designer, are a member of the organization for which an information system is scheduled. In this case you will figure in your own analysis; you are part of the **problem context**.

In our situation—a situation that often prevails—irrespective of any reductionist, 'hard', objective planning tools that we might use later on, an overall systemic, 'action research' (point 8 on the axis in Fig. 2.1) approach is essential. Figure 2.4 overleaf demonstrates the major components of an action research approach as set against the situation that can prevail if the approach is wrongly applied or not applied (anti-action research).

The action research approach (Fig. 2.4a) shows several important and useful themes:

1. The analyst and the client for whom the system is being designed work together as a team.

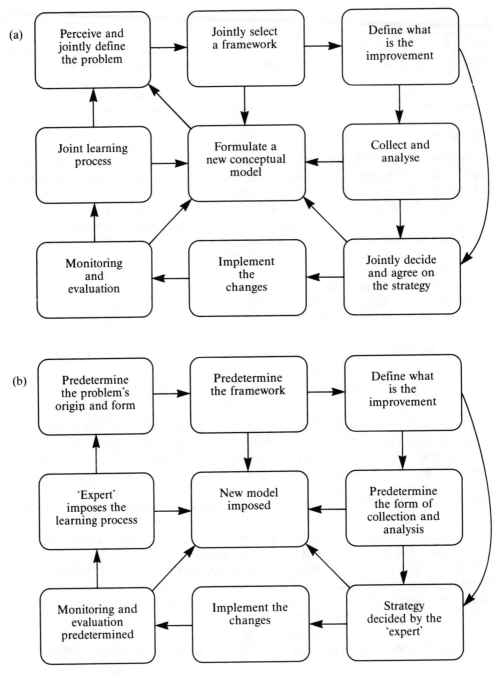

Figure 2.4 (a) Action research model. (b) Anti-action research model. (Adapted from Warmington, 1980.)

2. Strategy is jointly agreed on.
3. Final policy is jointly undertaken.

While not wishing to imply that any other approach would be fatally flawed, the anti-action research model (Fig. 2.4b) shows some of the range of problems that can arise if less attention is paid to the client.

The methodology that we will be using in the book is based upon this approach. By this means we intend to encourage planners to draw in interested parties to the work they are involved in, thereby reducing the possibilities of alienating stakeholders and/or missing vital organizational constraints which lie outside the narrow confines of the proposed information system. Further, the approach will allow us to see the way in which we as planners fit into the system we are devising.

The second question is that relating to methodology. The methodology we use here goes by the rather grand title of *a multi-perspective*, **eclectic** *methodology* evolved from field work based on 'Multiview'. If you think that this sounds off-putting perhaps we should explain that the approach is difficult to define without the use of terms like these but is much easier to understand and apply. What the title means is that the methodology makes use of a wide range of tools (it is eclectic) and attempts to perceive the problem that an information system confronts from a number of different directions (it is multi-perspective). The methodology consists of five components or tools. Four of these relate to methodologies that we have already discussed. Figure 2.5 shows the way in which these four relate to each other and to wider issues.

Figure 2.6 shows that two of the approaches are largely systemic and two reductionist. Also, two tend to be more concerned with the organization, while two are centred on the technology. The fifth component of our methodology is that which deals with the interface between the user and the computer itself, the **human–computer interface**.

We will go into greater depth concerning the specific details of the methodology in Chapter 4, but it is useful to see the overall layout now, and this is shown in Fig. 2.6.

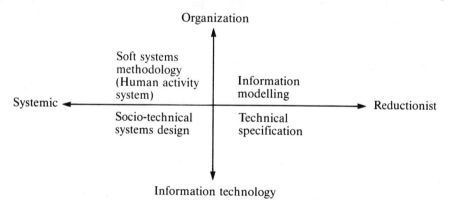

Figure 2.5 The four methodologies.

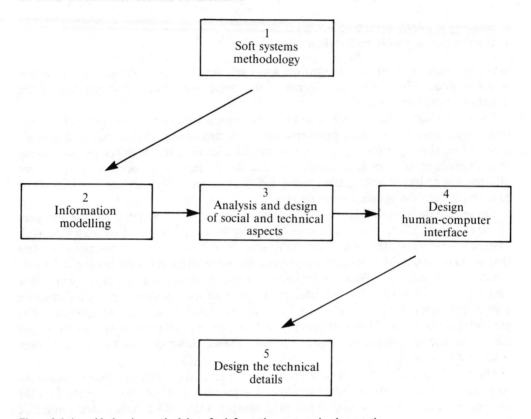

Figure 2.6 A rapid planning methodology for information systems implementation.

Before we discuss this methodology in depth we must define the role of the analyst (i.e. you).

2.3 CONCLUSIONS

In this chapter we have identified the approach to information systems planning that we are going to adopt and have demonstrated our reasons for selecting this approach in particular. Key points to remember are:

1. Our aproach will involve the active assistance of the recipient community.
2. The approach is not intended to be confrontational to any other but adopts:
   (a) key ideas from methods designed to improve the social significance of information systems and information technology;

Research type	Methodology	Problem-solving tools
Action research	Multi-perspective	An eclectic mix of both soft and hard

Figure 2.7 Systems analysis and systems design approach.

    (b) key ideas from technically rigorous methods which produce well-designed, technocratic systems.
3. We can therefore state our approach to have the characteristics shown in Fig. 2.7.

FURTHER READING

Antill, L. and Wood-Harper, A. T. (1985) *Systems Analysis*, Made Simple, Heinemann, London.

Argyris, C. (1985) Making knowledge more relevant to practice: maps for action. In E. Lawler *et al.* (eds), *Doing Research that is Useful for Theory and Practice*, Jossey Bass, San Francisco, Calif.

Avison, D. E. and Wood-Harper, A. T. (1990) *Multiview: An Exploration in Systems Development*, Blackwell Scientific, Oxford.

Checkland, P. B. (1985) From optimism to learning: a development of systems thinking for the 1990s. *Journal of the Operational Research Society*, **36**, no.9, pp. 757–767.

Checkland, P. B. (1988) Information systems and systems thinking: time to unite? *Interactional Journal of Information Management*, no.8, pp. 239–248.

Feyerabend, P. (1988) *Against Method*, Verso, London.

Haynes, M. (1989) A participative application of soft systems methodology: an action research project concerned with formulating an outlie design for a learning centre in ICI chemicals and polymers. M.Sc. thesis, University of Lancaster.

Klein, M., Heimann, P. and Money-Kyrce, R. E. (1955) *New Directions in Psycho-Analysis*, Tavistock, London.

Kozar, K. A. (1989) *Humanized Information Systems Analysis and Design: People Building Systems for People*, McGraw-Hill, New York.

Lucas, H. C. (1985) *The Analysis, Design and Implementation of Information Systems*, McGraw-Hill, New York. *Policy Sciences*, **14**, pp. 23–29.

Morgan, G. (1986) *Images of Organization*, Sage, London.

Winograd, T. and Flores, F. (1986) *Understanding Computers and Cognition*, Ablex, London.

Wood-Harper, A. T. (1989) Comparison of information systems definition methodologies: an action research, Multiview perspective. Ph.D. thesis, School of Information Systems, University of East Anglia, Norwich.

3

THE ROLE OF THE SYSTEMS PLANNER OR SYSTEMS ANALYST

Keywords the function of the analyst, past experience, methodology, area of use, self-analysis.

Summary The role of the analyst is to help the end users of the information system clarify their information-processing requirements and choose the most suitable systems design to meet these requirements. The analyst must perform the detailed analysis and work with programmers and others to help to implement a working system. This role, or some part of it, may be carried out from different positions within the organization, or from outside it. This section looks at how the analyst can arrive at a clear idea of his or her own background.

3.1 YOU THE ANALYST—1

All analysis must start from the basis that *reality is complex*. Information systems, intimately linked to so many elements of the social, technical, political, and cultural aspects of our lives, are also very complex. What we often fail to recognize fully is that we, as analysts, are also part of the overall context within which our information system will work. As our action research approach recognizes, we are within the research frame and we will influence

24

what goes on. Our own personal preferences will have an impact upon our planning and we will, consciously or unconsciously, attempt to influence stakeholders towards our own pre-set ideas about what is 'right' (as we saw in problem 2 in Chapter 1). This cannot be avoided and therefore is best understood at the outset.

It is not the purpose of this book to set out the complex and often confusing aspects of behavioural psychology and self-analysis. In this chapter we wish to make clear that our perceptions change over time, that these changes can be monitored, and that the understanding of our own personal bias will help us to understand the decisions that we make.

First of all, what is it that we as analysts and designers are trying to do?

Our first task is to attempt to make generalized models of the existing situation in order to create an information system. By formulating generalizations about current practices in organizations we can develop models of reality that we can then test for adequacy in hypothetical situations (e.g. 'Will this model pay-roll system cope with 23 staff being re-employed following dismissal notices being sent out accidentally?'). If our model is proved by experience to be adequate, then we can, with humility (i.e. recognizing that the system will always contain some faults and thus can always be improved upon), plan the working automated (or non-automated) system.

Understanding the complexity of reality is the nub of the analyst's dilemma with regards to making a reasonable model. Before going on to look at the tools the analyst employs, we need to consider the role that the systems analyst plays in an organization. Let us start with a very general definition:

> The systems analyst works with the user within his or her sociopolitical and economic context to specify the information system requirements of an organization. The system is modelled according to terms of reference and the final outline plans are produced for hardware, software and necessary processing.

This conveys the intermediary, 'go-between' aspect, as well as the architectural aspect of the job.

The title systems analysis and systems design is often used to convey the creative aspect of the role. There is a sense in which the analyst is like an architect producing designs to the clients' specification, or for their approval, which can then be turned into an actual construction by the programmer/builder, though this view minimizes the importance of the final system user in the analysis and design process. Our focus here is to set out:

1. How different types of individual conform to different types of analysis and design stereotypes.
2. How the recent history of analysis and design indicates how these stereotypes arose.
3. How a quick review of one's own intellectual background, methodologies, and work environment helps in assessing how our current approach has arisen.

First, and generally speaking, we identify four types or categories of analyst. The technocratic analyst (Fig. 3.1) fixes problems. He or she is best thought of as a technical expert like a doctor. The tendency of this approach is to take over the situation and impose one's 'expert advice'.

Self-image: a technical expert

Seeks: to 'fix' a problem with
objectivity and rigour

Figure 3.1 The technocratic analyst.

Self-image: agent for social
change

Seeks: to change radically the
existing status quo

Figure 3.2 The radical analyst.

In our second example the radical analyst (Fig. 3.2) seeks to overthrow existing wrongs and bring in new and improved systems. The metaphor of a warrior might be a bit strong but the underlying tendency of wishing radically to alter what currently exists is a fundamental aspect of the resulting approach. Here the analyst will attempt to assert the radical reform of current practices.

The third image (Fig. 3.3) is the one that most would wish to be associated with. Here the analyst seeks meaning and attempts to assist clients by facilitating their own problem-solving efforts. The analyst attempts to draw the clients into the problem-solving process and encourages them to become involved in all stages.

In our fourth example (Fig. 3.4) the analyst is again an agent of change but now in the sense of an emancipator—a catalyst assisting others to change their own lives. The difference between this analyst and the teacher is that here change and confrontation are inevitable. Therefore, the approach is 'hands-on' and can be highly assertive.

All four types of analysts can be seen on two axes as shown in Fig. 3.5.

The self-image of the analyst is very important. The four options we see here are derived from four very different perspectives: functionalist and interpretive at the top and radical at the bottom. Similarly the analyst can be seen as moving between the points of objective, 'scientific' behaviour and that of subjective preference.

To ask onself the questions:

- Am I seeking change or meaning?
- Am I planning a technical fix to the problem in hand or am I assisting others to recognize existing problems and develop internal solutions?

Self-image: a facilitator to assist
with problem solving

Seeks: the meaning of problems

Figure 3.3 The teacher analyst.

Self-image: a catalyst of change

Seeks: to change states of mind and
behaviour

Figure 3.4 The emancipator analyst.

- What is my self-perception?
- How will this affect the way in which I carry out my analysis and design?

is a useful precursor to undertaking research. Of course, any action arising from answering such questions is largely dependent upon ourselves. But it may well be that if one approach to analysis and design does not seem to be working, a different one can be tried. The answer to these questions largely depends upon where we see our own point of origin in the

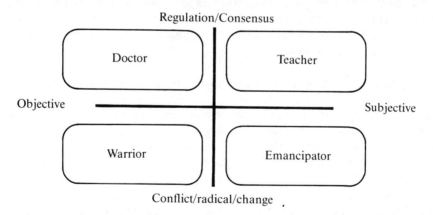

Figure 3.5 Four roles of the planner.

development of the present situation. To understand this it is useful briefly to look at the history of the analyst and designer.

3.2 THE HISTORY OF THE ANALYST

Others (e.g. Awad, 1990) have attempted to break down the recent history of computer development into distinct epoches. Here we take an adaptation of such a model and apply the salient features of the epoch as they concern the planner/analyst.

In the early days of **isolated computing** the hardware system was all important; the main use of computing systems was cost reduction accounting. The salient features of the period were:

- Mainframe-orientated systems.
- Computer experts were extremely remote (in terms of physical location and general attitude) from users.

The analyst was invisible in the system and dealt with technical matters.

Then followed **consolidated computing**, which was largely composed of programmers and poor documentation! The salient features were:

- Mainframe- and **minicomputer**-orientated systems.
- Programmers—no user interface!

Analysts, or more properly, programmer-analysts, were still buried in the system. They were mainly involved with designing systems for computers, not systems for people.

When the power of technical systems began to be appreciated by a wider managerial audience, faults arising from systems orientated to the wishes of programmers were recognized. At this time **management controls and restraints** were imposed. The key to this process was the enforcement of standards in terms of programming, new systems development, and system functions. The key features were:

- Mainframe-, minicomputer and microcomputer-orientated systems (micros were still disliked and considered a 'blip').
- Computer experts working directly under management control.
- Beginnings of crude user interfaces.

The analyst became a management aid. The beginning of the 'humanization' of many of these individuals began. Problems became organizational and less machine based (not 'what can the machine do?' but 'what can the machine do for me?').

This approach still did not allow easy access to computer power. The development of microcomputers has ushered in the (much celebrated) **role of the user**—focusing on applications software (e.g. packages) and distributed computing to remote officers (desktop micros), far from the computer department. Key features included:

- The advent of microcomputer networks.
- Experts acting as *facilitators* of user needs.
- Strict control of the computer function by the organization.

The analyst becomes central to understanding the needs of the user.

Most recently, following the trend above, we have seen the focus on **user–machine interface**, making users and computers more equal to the struggle of communication. Key notes for this process have been the development of user-friendly approaches.

Information technology becomes the norm in all parts of the organization.

- Invisible technology—'I don't want to understand it, I just want to use it'.
- Invisible experts—'Don't get in my way, just make the system easy to use'. We are all users now.

The theme here is that information systems are generally becoming more available to the user and are losing their technical/programming appearance. The analyst is now concerned with understanding users and making information systems dovetail into their needs.

The movement from isolated computing to the user–machine interface (which has a taken a mere 20 years) can be summed up by two statements which convey the focus of epoch 1 and epoch 5.

Epoch 1 The computer expert is the centre of the system. The computer expert is given the necessary support to indicate priorities and to control the process of providing automated procedures to alleviate problems. The user is peripheral to the needs of the data-processing department and acts as a problem object to the computer expert.

Epoch 5 The user is the centre of the system. The user is given the necessary support to indicate priorities and to begin the process of providing automated procedures to alleviate problems. The computer expert is peripheral and acts as counsel and support to the user.

Perhaps the most striking point that arises from these two statements is the movement from a hard, reductionist and 'scientific' view to one that focuses on the needs of the often despised user.

3.3 YOU THE ANALYST—2

Because background determines research approach, influences methodology, and determines problem-solving tools, it is quite useful (though still largely not practised by analysts and designers) to review one's own background with regard to the particular situation in which you are going to work. The analyst is as human as anyone else, and if you intend to carry out your own analysis it is useful to have a system to recognize, before beginning the process of analysis, where your own ideas and concepts arise, and to consider how likely they are to influence the task in hand.

3.3.1 Present self-analysis

One means for such self-analysis is shown in Fig. 3.6, which demonstrates a fairly crude but easy-to-use tool for identifying the predispositions of the analyst with regard to background, immediate problem context, and methodology being used.

> If you have never undertaken analysis and design before you should still be able to express a methodology preference from the information given in Chapter 2.

The first object is to identify background. This ranges from social studies to engineering, from soft to hard, from systemic to reductionist. The second question—with regard to methodology—similarly sets the task of identifying a soft or hard approach. The third question indicates the technological sophistication of the area being worked in (e.g. a computer-wise city bank or a naïve farmers' cooperative), and the fourth question asks for an indication of risk and uncertainty in that environment.

The result will be a mark between 0 and 20 for each of questions 1–4. For questions 1 and 2, marks tending to 0 indicate a soft background. Marks tending to 20 indicate hard. For questions 3 and 4, marks tending to 0 indicate high risk and low sophistication, marks tending to 20 indicate low risk, high sophistication.

A useful rule of thumb for understanding this type of exercise is shown in Fig. 3.7 overleaf. The tool can be used as a rapid way of assessing the appropriateness of the analyst's approach in the given situation.

3.3.2 The history of self-development

This is another approach, which can be used in conjunction with that set out in Figs 3.6 and 3.7. The method is intended for those who have undertaken systems analysis and systems design before and consists in reviewing personal development over several years. This can be undertaken by focusing on three key areas (see Fig. 3.8 on page 32):

1. What is my intellectual framework? That is, the set of ideas and principles that underlie the way I work.
2. What methodologies have I applied? If you have not undertaken systems analysis and systems design before, what is your manner or style of working?
3. What were the situations in which I worked?

Look at your own intellectual framework as thoughtfully and impartially as possible. It

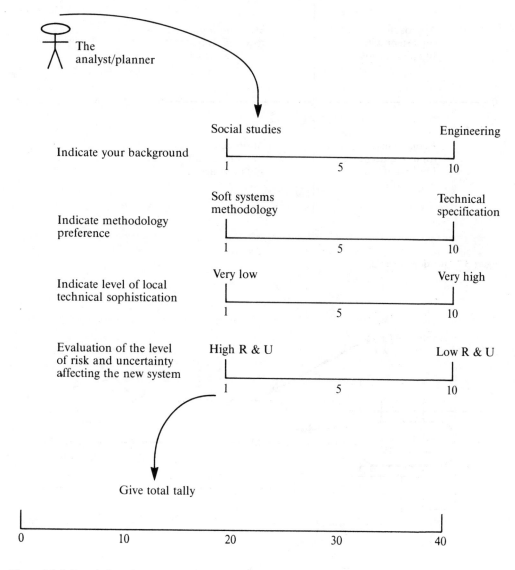

Figure 3.6 Self analysis tools—'analyst know thyself'.

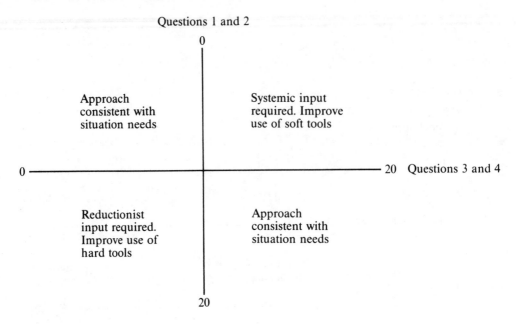

Figure 3.7 Questions 1 and 2.

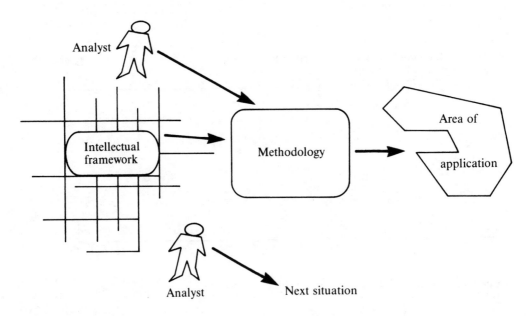

Figure 3.8 Previous experience.

will tell you a lot about how you will approach the subject of the analysis (warrior or doctor?) and may also indicate whether you will tend towards technical solutions or social measures to make systems work. 'Man know thyself' is as relevant to the system analyst as it is to the mystic. One example of such an analysis is shown in Fig. 3.9 on page 34. The figure shows five major shifts in the intellectual background, methodology, and area of application of an analyst. It generally indicates a movement from hard to soft approaches. Figure 3.9 shows five 'snapshots' in an individual's development. The example is academic but gives some general themes that are of interest. The snapshot shows five frames over five years. Intellectually the analyst moves from his university concentration on 'development studies' (the study of third-world development), through the reductionist school of analysis and design, to soft systems approaches like **Web** and **Multiview**. His methodological development mirrors this movement, from hard technical specification to soft multiperspective. Areas of application range from East and West Africa to Asia.

A more business-orientated approach is shown in Fig. 3.10 (see page 35). This second model, which does not relate directly to information systems design, shows the key snapshots in the development of an accountant. She begins work in a small electronics company and works to the book of her accountancy training. The second frame shows her movement to local government. A team approach is important in a large company and with this comes what we call a corporate mentality. The mentality requires the accountant to concentrate on the aspect of the local government body she is working for and not the total situation —her experience when working for the electronics company. The third frame shows a switch in the direction of local government (such as that which occurred in the UK in the 1980s). Accountancy thinking becomes much more 'predatory'. Costs are being cut and fiddles are being sought. This in turn brings a competitive edge to team work and also introduces a 'dynamic' edge to the team concept.

3.4 CONCLUSIONS

As we have seen in Chapter 2, systems analysis and systems design is a highly complex subject incorporating many different 'flavours'—from the hard and scientific to the soft and social sciences orientated.

Along with the analysis of problem contexts we have to recognize that we, the analysts and designers, also come into the frame. We affect what we work upon. In order to be fully aware of the impact we are having upon the work in hand it is quite useful to know:

1. Where our own strengths and weaknesses are in the problem context. We can get this information from a 'Present self-analysis'. The task in hand may require a shift in our present approach.
2. What our path has been to the present situation, and therefore what our overall tendencies have been in terms of intellectual development and work methodology. We can ascertain this information from 'the history of self-development'. Our current task may require a substantial change in intellectual framework or methodology.

The analyst has:

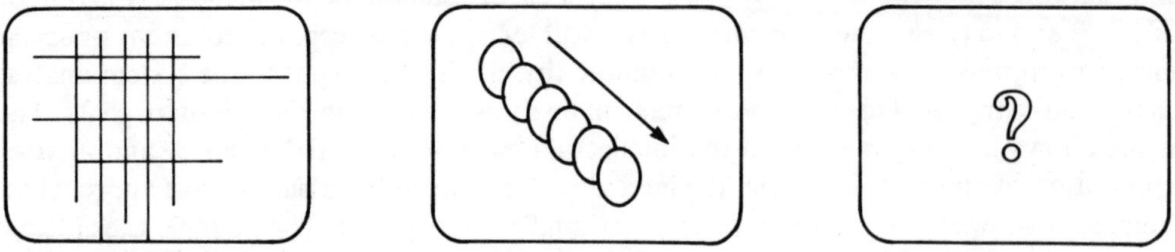

| an intellectual framework, | a methodology, and | an area of application (or problem context) |

Frame 1

| Development studies literature | Technical specification (TS) | East Africa |

Frame 2

| Hard systems analysis | Amended TS | West Africa |

Frame 3

| Critical reductionist, 'Muddling along' | Second amended TS | West Africa |

Frame 4

| Web model Soft systems MultiView | MultiView | Asia |

Frame 5

| Web model Soft systems Multi-perspective | Multi-perspective | Africa |

Figure 3.9 The chronological development of an analyst.

The accountant has:

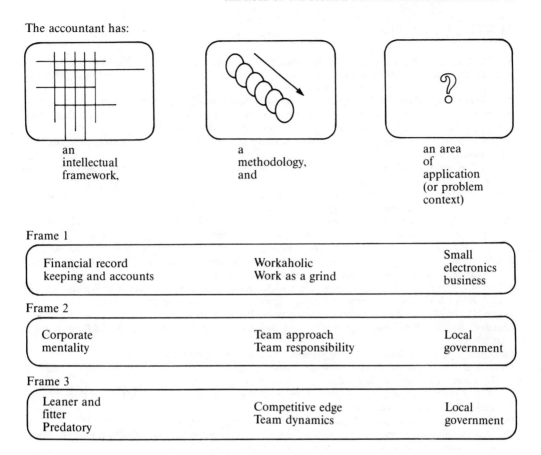

Figure 3.10 The chronological development of an accountant.

These two self-analysis procedures provide the analyst with an overview of his or her current state. This can be vital if problems arise with the analysis and design procedure and there is a need to rethink the approach. For example, there may be a situation in which the analyst favours the soft approaches, tending to focus on user self-help and workshops to discuss problems. The client requires more leadership and drive from the analyst ('why ask us to think it out, that's what we are paying you to do'). This requires the analyst to shift focus to the hard tools in the methodology and to adopt a more managerial syle.

As we shall see in Chapter 4, the methodology we apply here allows the analyst a certain amount of freedom in the selection of tools in the problem context. If problems arise due to the analyst's approach, the analyst can substitute soft for hard tools (or vice versa) or can reschedule their use.

FURTHER READING

Ad Hoc Panel (1987) *Microcomputers and their applications for Developing Countries*, Westview, Boulder, Colo.

Anthony, R. N. (1965) Planning and control systems: a framework for analysis. Harvard University Graduate School of Business Administration, Boston, Mass.

Antill, L. and Wood-Harper, A. T. (1985) *Systems Analysis*, Made Simple Computer Books, Heinemann, London.

Argyris, C. (1985) Making knowledge more relevant to practice: maps for action. In E. Lawler *et al.* (eds), *Doing Research that is Useful for Theory and Practice*, Jossey Bass, San Francisco, Calif.

Avison, D. E. and Wood-Harper, A. T. (1990) *Multiview: An Exploration in Systems Development*, Blackwell Scientific, Oxford.

Avison, D. E. and Wood-Harper, A. T. (1991) Information systems development research: an exploration of ideas in practice. *Computer Journal*, **34**, no.2.

Bell, S. (1986) Information systems planning and operation in less developed countries, three parts. *Journal of Information Science*, **12**, nos 5 and 6, pp. 231–245, 319–331 and 333–335.

Bell, S. (1987) A guide to computing systems evaluation and adoption for users in LDCs—some problems encountered in applying standard techniques. *Journal of Information Technology for Development*, **3**, March.

Bell, S. (1990) The information technology 'Fix—lessons from the third world. *Computers in Africa*, forthcoming.

Bell, S. and Shephard, I. (1990) Increasing computerisation in the 'new democracies' of Eastern Europe—lessons from the third world. Paper prepared for a conference in Gdansk.

Boland, R. (1985) Phenomenology: a referred approach to research in information systems. In F. Mumford *et al.* (eds), *Research Methods in Information Systems*, North Holland, Amsterdam, pp. 193–201.

Bowers, D. (1988) *From Data to Database*, Van Nostrand Reinhold, London.

Chambers, R. (1981) Rapid rural appraisal: rationale and repertoire. *Public Administration and Development*, **1**, pp. 95–106.

Checkland, P. B. (1983) *Systems Thinking, Systems Practice*, Wiley, Chichester.

Davis, G. B. (1984) Systems analysis and design: a research strategy macro analysis. Mimeo, Carlson School of Management, University of Minnesota.

Davies, L. J. (1989) Cultural aspects of intervention with soft systems methodology, Ph.D. thesis, Department of Systems, University of Lancaster.

Douglas, J. D. (1976) *Investigative Social Research*, Sage, Beverly Hills, Calif.

Gorry, G. A. and Scott-Morton, M. S. (1974) A framework for management information systems. In R. Nolan, (ed.), *Managing the Data Resource Function*, West Publishing Co., St. Paul.

Gotsch, C. (1985) Application of microcomputers in third world organisations. Food Research Institute, Stanford University Working Paper, no.2.

Grant-Lewis, S. (1987) Computer diffusion in Tanzania and the rise of a professional elite. Paper presented at the African Studies Association Annual Meeting, Denver.

Han, C. K. and Walsham, G. (1989) Public policy and information systems in government: a mixed level analysis of computerisation. Management Studies Research Paper no.3/89, Cambridge University Engineering Department.

Hirschheim, R. (1984) Information systems epistemology: an historical perspective. In E. Mumford, *et al.* (eds), *Research Methods in Information Systems*, North Holland, Amsterdam.

Kling, R. (1987) Defining the boundaries of computing across complex organisations. In Boland, R. J. and Hirschheim, R. A. (eds), *Critical Issues in Information Systems Research*, Wiley, Chichester.

Kling, R. and Scacchi, W. (1982) The web of computing—computing technology as social organisation. *Advances in Computers*, **21**.

Kozar, K. A. (1989) *Humanized Information Systems Analysis and Design: People Building Systems for People*, McGraw-Hill, New York.

Land, F. (1987) Is an information theory enough? In Avison *et al.* (eds), *Information Systems in the 1990s: Book 1—Concepts and Methodologies*, AFM Exploratory Series no.16, Armidale NSW, University of New England, pp. 67–76.

Lucas, H. C. (1985) *The Analysis, Design and Implementation of Information Systems*, McGraw-Hill, New York.

Mumford, E. (1981) Participative system design: structure and method. *Systems, Objectives, Solutions*, 1, pp. 5–19.

Nolan, R. L. (1984) Managing the advanced stages of computer technology: key research issues. In W. McFarlan, (ed.), *The Information Systems Research Challenge*, Harvard University Press, Boston, Mass.

Tagg, C. and Brown, J. (1989) TBSD: notes on an evolving methodology. Computer Science Technical Note, School of Information Sciences, Hatfield Polytechnic, UK.

Wood-Harper, A. T. (1990) Comparison of information systems approaches: an action-research, Multiview perspective. Ph.D. thesis, University of East Anglia, Norwich.

4

TERMS OF REFERENCE AND SELECTING OUR PLANNING/DEVELOPMENT TOOLS—SEQUENCE AND SCHEDULE

Keywords project cycle, terms of reference, human activity system, root definition, rich picture, conceptual model, information modelling, social and technical systems design, human–computer interface, technical subsystems, tool selection, context.

Summary All planning or analysis and design begins with a set of terms of reference. Following these, the analyst will have some idea as to what specific work is expected, under what conditions, and with what resources. Following on from this, the analyst can select the tools that are appropriate within the context of the problem being reviewed and set out their sequence and schedule.

4.1 THE REALITY OF ANALYSIS: TERMS OF REFERENCE

It would be pleasing to the ego and satisfying to the power hungry to believe that the analyst can be all-powerful in the problem context. Like Caesar, to cry 'Veni, vidi, vici' (I came, I saw, I conquered) would be a rather satisfactory way of concluding the analysis. This will not happen to you very often if common experience is anything to go by!

Many systems analysis and systems design books set out as though the analyst's word was law and the specified logic of the analysis was always carried out to the letter. This is rarely the case and possibly especially so in situations of rapid change and risk. Financing agencies, be they banks or accounting departments, putting up the cash for analysis and design, tend to impose very strict guidelines or *terms of reference* upon the analyst, which will mean that a certain amount of prejudging of the situation will have taken place (sometimes by individuals carrying out feasibility studies with little knowledge of information technology, sometimes by managers who think that they already know the answer to the problem).

It is no good the analyst specifying a new, minicomputer-based management information system for an organization if the terms of reference restrict all further development to simple microcomputer database functions. Sometimes you may feel disheartened when your analysis tells you very different things from the guidelines you have received.

The ability of the analyst to move freely within the context of his or her terms of reference and the associated budget and labour limits will depend very much upon:

- The ability of the analyst to convince the funding body that more or less may be required (the latter is easier!).
- The willingness of the funding body to be flexible.

The golden rule is never to exceed the boundaries of the system as seen by the funding body without first convincing all the major stakeholders in the system that such a course is both right and necessary.

This introduces a larger issue, namely the position of the analysis and design procedure in what is called the project cycle (see Fig. 4.1 overleaf).

Analysis and design is usually just one aspect of a total project, e.g. a project for the development of a new factory and office site in the north-east of the United Kingdom. One issue within this project is the information system required for organization support.

Your analysis and design may well have to fit into such an overall cycle. This will require team work and compromise on key issues. If you wish to have more details on the project cycle, see Appendix 2.

4.2 THE CONTEXT OF AN ANALYSIS METHODOLOGY—SELECTING THE RIGHT TOOLS

The first activity within our analysis is to select the tools that are appropriate to the situation under study, i.e. those that conform to:

1. The conditions set out in the terms of reference.
2. The personal preferences set out in our self-analysis (outlined in Chapter 3).

We have already introduced, in outline only, the methodology tools that we are going to

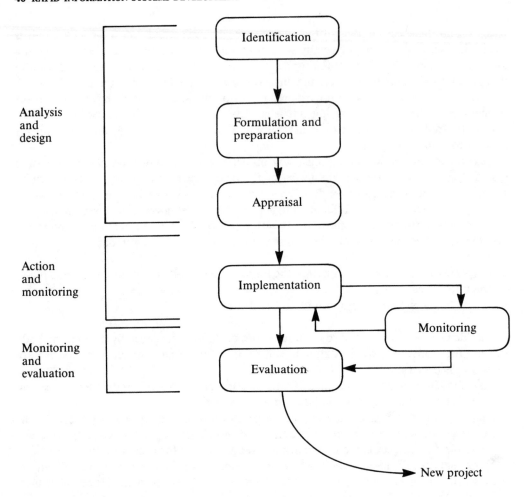

Figure 4.1 The project cycle. (Adapted from Coleman, 1987.)

set out in this text. These are shown in Fig. 4.2 opposite. We now need to flesh out what these tools actually do.

4.2.1 Soft systems methodology

This comprises the analysis of what Checkland has defined as the **'human activity system'** (HAS). The HAS is the concession of our approach to soft systems methodology (SSM) and is in turn composed of three core items:

- The **rich picture**, which is devised to show the principal human, social, and cultural activities at work in the perceived environment. The rich picture usually includes the

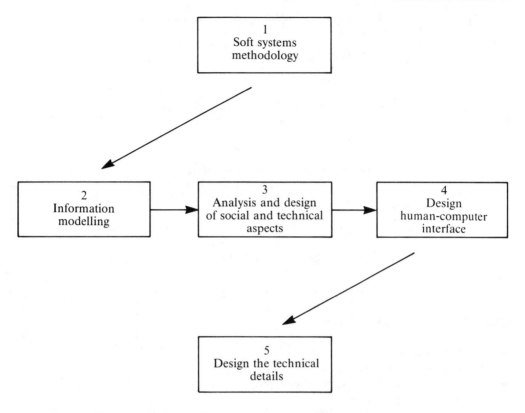

Figure 4.2 A rapid planning methodology for information systems implementation.

structures and processes at work in an organization.
- The **root definition**, which by identifying the key clients, actors, things to be done, assumptions, problem owners, and environments, attempts to structure the results of the rich picture analysis into a mutual (analyst and stakeholders) perception of 'what can we do about the problem?'
- The **conceptual model**. This model is used in our approach as a simplification of the key tasks and areas involved in the new improved information system. The conceptual model is an outline of what we are going to attempt to design.

4.2.2 Information modelling

The second phase of the analysis is **information modelling**. In this phase we adopt a more reductionist and technical approach. At this stage we want to develop the conceptual model, which by definition is an idea requiring structuring into a workable system. In information modelling we attempt to draw together:

An **Entity** might be – an employee

the entity has

Attributes — date of birth, date of joining
grade, etc.

the entity carries out

Functions — produces annual staff report,
annual grading changes, random
promotions, new staff enrolment,
staff loss, etc.

at particular

Events — end of year, staff enrolment, etc.

Figure 4.3 Example of an information model.

- the major **entities,**
- the **functions** of these entities,
- the **events** that produce these functions, and
- the **attributes** of the entities.

The simple application of this tool is to be able to generalize the key points identified in the conceptual model down to a set of human and computer-related tasks and objects that can be designed into a new system. Figure 4.3 is a simple example of an information-modelling exercise in overview.

4.2.3 Social and technical requirements

The third phase requires that the analyst bring together the right mix of **social** (human resources) and **technical** (information technology, other technology) **requirements**. Here the key hardware and identified human alternatives, costs, availability, and constraints are married together. This stage produces the right mix of technology and manpower to implement the system outlined in phase 2.

4.2.4 Human–computer interface

The fourth phase deals with the human–computer interface (HCI). This involves thinking about the means by which the two aspects of the proposed information system (human beings and technology) can best communicate with each other.

4.2.5 Technical aspects

The fifth and last aspect involves the design of the necessary technical aspects that combine to produce the overall technical design. The major technical subsystems are shown in Fig. 4.4. The six major elements shown are arguably the core of any information system:

- The *applications* area deals with transactions within the computer (inputting data, updating records, gathering data elements for print out).
- *Retrieval* deals with the output from the system.
- *Database* is the core structure containing entities of the computer system.
- *Maintenance* deals with both preventative and corrective maintenance.
- *Management* controls the overall information system process within the organization context.
- *Monitoring and evaluation* deals with the effective performance of the information system.

The current task is to select which, if not all, of the tools to use. The methodology as a whole

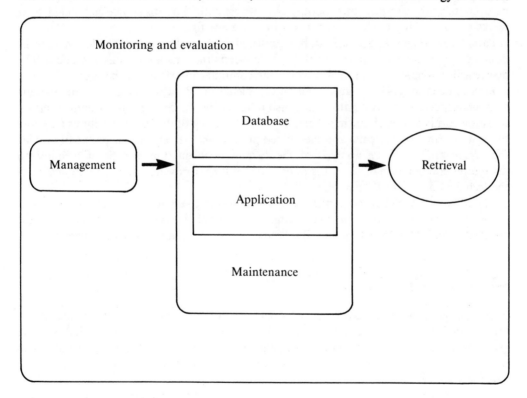

Figure 4.4 Monitoring and evaluation.

can be set out as shown in Fig. 4.5. Again, do not panic about the complexity of this picture. Figure 4.5 gives us some more detail. First it is worth noting that the first stage of the analysis involves an iterative loop, or period of discussion and feedback between the analyst and the stakeholder group. The second point to note is that the second stage of the analysis, information modelling, may throw up inconsistencies that we will need to rethink. This type of eventuality is impossible to foresee. An example would be where the conceptual model requires two departments to share one common information product such as salary details but where in fact this idea is strongly objected to by staff. This would require a reworking of the model and possibly the rich picture. The third point to note is that following the fifth stage of the process comes software selection, hardware selection, and implementation strategy. These issues are not strictly part of the analysis and design, but general issues which will be dealt with in Chapter 10.

The major constraint on the use of tools is cost and time. With analysis and design this simplifies down to time. If we return to the overall picture of our methodology we can identify three separate ways in which the tools can be used (see Fig. 4.6 on page 46).

Obviously the three paths offer three different levels of analysis. Path 1, the five-stage path, contains the complete methodology and we estimate that this can be completed in six weeks or 30 working days, though this is only a guide figure.

Path 2, four stages, as a guide can be completed in 25 working days. The loss here is the design of the human–computer interface. This means that the analysis as a whole will be deficient in planning the manner in which the computer interfaces to the user.

Path 3, three stages, as a guide can be completed in 18 days. The loss is the human–computer interface (as with path 2) and also information modelling. This is quite a serious omission and it will result in there being no clear planning of database structures (the core of most information systems) and the related programmes. Even so, this path will provide the organization with an analysis of the current problem, the best fit of technology/manpower combination of resources to provide for the improved system, and an outline of the technical details of that system.

The following five chapters outline each of the stages given above. Your current task is to select the best path for your specific analysis. Your choice needs to be based upon the constraints of budget and time and the needs for a detailed analysis and design.

4.3 CONCLUSIONS

Following on from the review of the approach we are adopting set out in Chapter 2 and the self-analysis set out in Chapter 3, this chapter requires us to select tools that we have the resources to use in our analysis and design. Before going on to the analysis and design proper, be sure in your own mind that you know which of the three paths you are going to adopt.

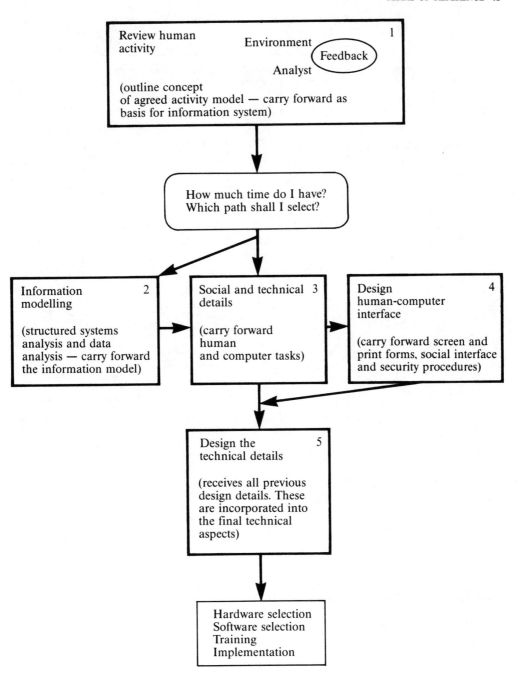

Figure 4.5 Rapid planning methodology.

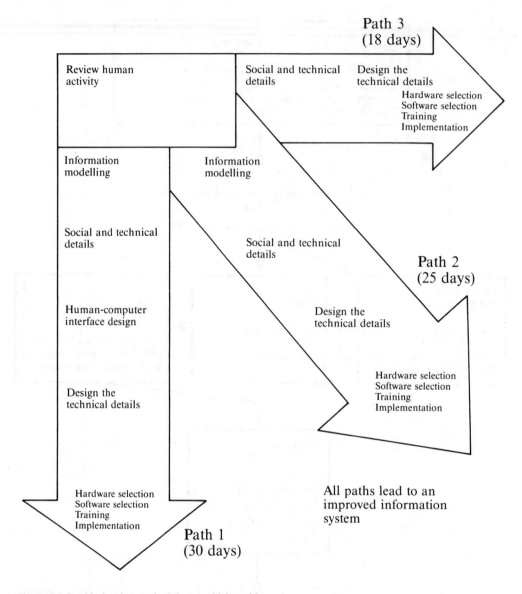

Figure 4.6 Rapid planning methodology—which path?

FURTHER READING

Avison, D. E. and Fitzgerald, G. (1988) *Information Systems Development: methodologies, techniques and tools*, Blackwell Scientific, Oxford.

Avison, D. E. and Wood-Harper, A. T. (1990) *'Multiview': and Exploration in Systems Development*, Blackwell Scientific, Oxford.

Avison, D. E. and Wood-Harper, A. T. (1991) Information systems development research: an exploration of ideas in practice. *Computer Journal*, **34**, no.2.

Checkland, P. B. (1983) *Systems Thinking, Systems Practice*, Wiley, Chichester.

Land. F. (1987) Is an information theory enough? In Avison *et al.* (eds), *Information Systems in the 1990s: Book 1—Concepts and Methodologies*, AFM Exploratory Series No 16. Armidale NSW, University of New England, pp. 67–76.

Lucas, H. C. (1985) *The Analysis Design and Implementation of Information Systems*, McGraw-Hill, New York.

Mumford, E. (1981) Participative system design: structure and method. *Systems, Objectives, Solutions*, **1**, pp. 5–19.

Winograd, T. and Flores, F. (1986) *Understanding Computers and Cognition*, Ablex, London.

5

WHAT IS THE PROBLEM? THE HUMAN ACTIVITY SYSTEM—MAKING A MODEL

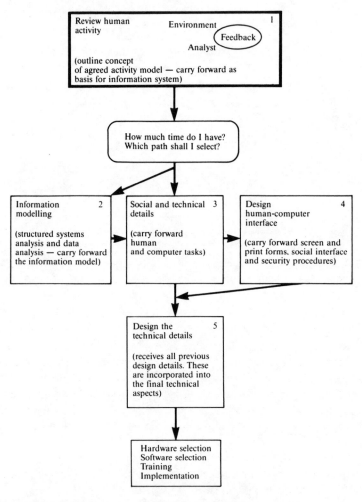

Figure 5.1 Understanding the environment—rapid planning methodology.

Keywords human activity system, rich picture, context of the user, root definition, new conceptual system.

Summary To produce a useful analysis and design the analyst must clearly define the key elements of the situation and relate these to the terms of reference. The chapter makes this connection and develops the analysis within the context of the rich picture, the prime means for understanding the context for the information system in terms of issues and tasks. Having agreed the picture, the major factors of the system environment are developed in terms of the root-definition (who is doing what for whom in what context) and agreed with the major stakeholders. Finally, a conceptual model is produced that gives the outline of the proposed new system; this can then be fed into the next stage of the analysis.

5.1 INTRODUCTION TO THE HUMAN ACTIVITY SYSTEM

It should be noted that the development of the human activity system, as set out in this chapter, is the result of practice in developing countries as opposed to other varieties of interpretation (see Checkland and Scholes, 1990, for the definitive version). In most contexts of information systems development there is a need and a problem (at least one of each!).

The perception of the problem situation and the resulting definition of the need for information is the nexus of this stage of analysis. Our job is to alleviate the problem by improving the information-processing capacity of the organization.

> We should not confuse the concept of the information system with the more specific *automated information system*. Quite often the situation requiring analysis already contains a manual information system and the result of all the analysis and design may be to *prescribe a revised manual system*.

With the key ideas of *need* and *problem situation* in mind, we can say that the problem as such exists within the context of the human activity system (HAS). The HAS can be seen as a view on the social/cultural/ethical/technical (etc.) situation of the organization. In outline we can see the process of HAS analysis as shown in Fig. 5.2 on the next page.

Figure 5.2 begins from the standpoint of perceiving the new information problem situation. From this arises our first tool, the creation of the structures of the rich picture. This should define for us the major tasks and processes involved in the problem context. The next stage of the analysis, and the next tool to be used, is the root definition, the CATWOE criteria (more on this later, but meaning: who is doing what for whom, under what assumptions, and in what environment), which sets out the fundamental features of the problem context. From this definition we can create a new model of the improved situation as we see it. This model is called a conceptual model.

Do not worry about these phrases here. For a definition of each see the Glossary. Each will be developed in the next thirty pages or so. Like most new and grand sounding phrases you will find them quite simple concepts in themselves.

For now, let us return to our problem context, the organization. This organization may

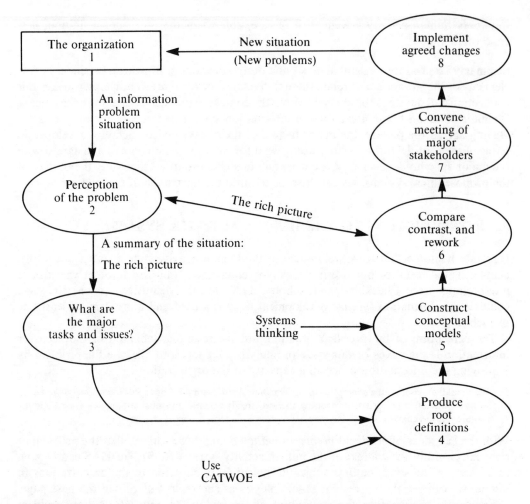

Figure 5.2 Overview of the soft systems methodology procedure. (Adapted from Wood-Harper, 1989, and Avison and Wood-Harper, 1990.)

be a research institute, a government department, a training centre, or a single office within some large organization. It may be one person, a sole researcher wanting to keep research records, or a teacher recording student progress.

The purpose of this phase is to assist the analyst because he or she must understand the HAS in order to study the information flows involved in the organization. With this in mind the first job of the analyst is to help the major stakeholders in the organization define the situation and analyse what the problem is so that they can set about solving it. The analysis of the situation as viewed as a HAS as shown in the Fig. 5.2 consists in:

1. Perceiving the organization's problem situation as defined in the terms of reference.
2. Identifying, by means of the rich picture, the tasks and issues.
3. Identifying and noting key conflicts of interest.
4. Coming to an agreed definition of the problem that is to be tackled.
5. Setting out the outline of the improved system themes in the conceptual model.

The following sections will develop this theme.

5.2 THE RICH PICTURE

5.2.1 Preparation

Often professional analysts will have very little knowledge or understanding about the range of issues involved in the target organization's information-processing work. This book is intended for non-professionals and for those planning information systems for their own organization, so we do not assume that you will necessarily be new to the problem context. It can be an advantage to have little knowledge of the organization in which you are working! An outside analyst will have terms of reference, timescale and budget, and an outline job to do. As an outsider the analyst will not have problems with existing staff relationships or subjective preferences concerning the way the organization is run. The analyst often needs to understand the problem context rapidly and in this process of understanding has a chance to bring a degree of *impartial and uncommitted* analysis. If, on the other hand, you are a member of the organization in which your analysis and design is taking place, you probably will have developed your own ideas, which will often be unstated and sometimes not consciously recognized. This can cause problems. For example, the analyst has a strongly held belief that department x is a better candidate for automation rather than department y. This belief is based upon close familiarity with the one department and comparative ignorance of the other. New systems that reflect this view may cause problems for the user community and ultimately for the analyst. Outsider analysts may also have problems—e.g. imposing their expert opinion on situations of which they have little understanding (see Anthill and Wood-Harper, 1985). To reduce the likely problems that may arise from this type of subjective preference we have a series of *analysis and design application tools* that we will use throughout the following analysis:

1. User/client (or stakeholder) participation is essential if the analysis is to be useful. The problem is not the analyst's property—it belongs to the organization and for this reason the individuals in the organization must be brought into partnership with the analyst designer as part of the problem-solving team. Tools for doing this include:
   (a) A preliminary meeting with all those concerned with the analysis and design—setting out and discussion of the terms of reference. The stakeholders should be

encouraged to comment on the task and to make any observations on the way in which the analysis and design might develop.

    (b) Regular workshops throughout the analysis and design for briefing and sharing of views.

2. Rigorous application of the agreed terms of reference. Many forms of systems analysis and systems design tend to spill over into areas that are not contained in the original problem. This is quite easy to do with systems work. Nevertheless the practical analyst must focus on the issues that are of primary concern. This does not mean that other areas are to be ignored. If the new system impinges upon a large area, recommendations can be made for a wider study at a later date.

3. Reporting. All stages of analysis need to be adequately reported, primarily for the analyst's own benefit and also as an aid for the stakeholders. Information systems professionals are renowned for providing poor or no documentation.

4. The use of interview techniques. Books have been written on the art of interviews. Key points for the analyst are:

    (a) Initial contact—dress and manner should be appropriate to the problem setting. It is surprising how many analysts 'lose' their object of study by appearing too glib, off-hand, or conceited.

    (b) Sequence—it is a good idea to lead in your interview with some light and non-threatening conversation (especially with those who seem most uncomfortable with the idea of information systems). More detailed questioning can then follow.

    (c) Questions must be understandable. This may appear obvious, but quite often computer-related questions are far from obvious to those being asked.

    (d) Caution is required when pushing into areas that are sensitive (internal audit, interdepartmental competition, etc.). A lack of tact can cause an interviewee to dry up.

    (e) Always be neutral in your style.

    (f) Again a basic point—be sure to document the interview, you *will* forget much of the detail otherwise.

5. Basic observation of site and behaviour. Many key factors for a successful analysis taking place in limited time are literally obvious. By keeping eyes and ears open we do not neglect the obvious (staff aggression, resentment, poor filing, shabby record keeping, etc.).

Quite often the analyst will discover that there has never been a prior analysis or review of the organization's information-processing problems and capacities and there may be a fair degree of surprise at some of the findings of the rich picture.

5.2.2 The primary components of the rich picture

Structures The way in which we produce rich pictures is composed of two elements—structure and process. These are divided into two key areas—technical 'facts' (hard areas) and social/ethical/cultural realities (soft areas). Throughout this book we will be working on one key example and several minor ones to explain the way in which the methodology works. Our first information will be a set of terms of reference. For the purpose of the example we are working out in this book these are as follows:

- Terms of reference—'outline non-specialist functions that could be computerized in order to increase efficiency and timeliness in terms of day-to-day operations of a small government department in a developing country.'
- Resources—one analyst.
- Time allocation—30 days.

You will need to consider your own terms of reference before proceeding. Our example is selected intentionally to show how systems analysis and systems design tools can be used in the most hostile situations (in terms of cross-cultural analysis, tough climate, low access to expert skills, etc.) and still produce useful information systems.

The easiest way to begin is to produce a map or cartoon of the major structures to be involved in the picture. These may be departmental boundaries, system boundaries, national borders, etc., as they are applicable to the problem in view. Working our way towards the eventual picture we begin the exercise by setting out the '**hard**' structures as shown in Fig. 5.3.

Immensely simple as this view is, it shows that the analyst is aware of various agencies at work both within and outside the focus of the analysis, which in this case is a governmental department. Our hard structure tells us that the department is composed of at least six discrete sections or areas of activity and that there are three key structures operational outside the boundary of the department that will impinge upon the eventual system to be set up. Note that there will be quite a number of other agencies and groups at work within the context of the department. This initial portrayal of the major areas of interest already shows that we are beginning to focus down on to what the analyst believes to be the key

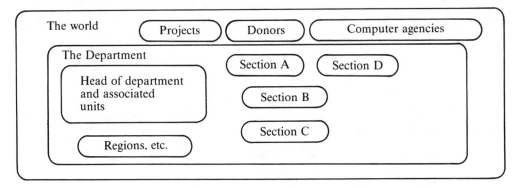

Figure 5.3 The rich picture. Major structures—'hard'. (We begin by depicting the major structures within the organization under study. Essentially we group the structures under two headings: soft and hard.)

areas of concern as expressed in the terms of reference—the major job of departmental management.

Our next task is to set out the less formal or **'soft'** structures active within the overall problem context. The important feature of this stage is to set out structures which, although we identify them as being essential for the eventual working of the system, are related more to cultural and ethical than technical points. Because of the sensitivity of some structures it may not always be possible to show these 'soft' structures in workshops with stakeholders and the final report. In our case the types of structure are shown in Fig. 5.4.

In the 'soft' picture we identify a range of structures, some more formal (the promotions committee), some more informal (language and cultural groups), and some political (faction and interest groups). The reason all these structures are termed 'soft' may not be immediately obvious, but is because they are identified in our analysis to date as having subjective and therefore to some extent unguessable effects. For example, the promotions committee in most normal circumstances would be seen as having a 'hard', objective identity in the department. In this case we are not interested in the committee's function and purpose; we are interested in one of its informal functions, which is to move staff around for political reasons. This could have a major impact on an embryonic computer or data-processing unit. Another 'soft' structure is the externally funded donor unit within the

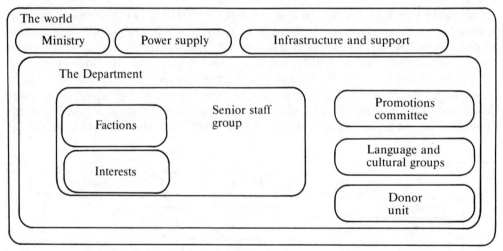

When is a structure 'hard' or 'soft'?

The analyst is the final judge.

Generally 'hard' are fixed and formal, 'soft' are variable and informal.

This is not always a 'scientific' process.

Figure 5.4 The rich picture. Major structures—'soft'.

department. Again it is a physical unit with a task to accomplish not directly related to the work in hand. However, its presence is felt by most major actors in the department and for this reason it has a subjective (true or untrue) watchdog function. Most obviously 'soft' are the cultural and language groupings in the department. Any incoming system has to work with the dominant theme in terms of culture and will have to reach an accommodation with other interests. Almost subliminal to the outside, short-term consultant, but vital to note, are the interest groups and factions within the senior staff groupings. These may not have a direct impact on the project as a whole but they do need to be understood and planned around. In the outside world we have a ministerial watchdog, problems of power supply failure and fluctuation, and the lack of infrastructural support (hardware and software support).

As already noted, it will not always be advisable to identify all structures in reports and workshops. There are often good working reasons why an analyst wishes to keep clear of unnecessary controversy. This is part of the reality of understanding analysis and design in context. The result of the construction of a rich picture should be the identification by the analyst of what is possible within the problem context. What is and is not said and made explicit is a decision left to the discretion of the individual analyst.

Processes Our next job is to identify 'hard' and 'soft' processes operating upon structures in terms of the overall work of the department. As above we can develop our thinking with two separate models. Figure 5.5 overleaf demonstrates the relationship between the structures and processes in the 'hard' context.

The processes that we set against each function are obviously only part of the whole range of activities performed. This demonstrates again the subjective nature of the analysis (there is never a 'right' rich picture) and the attempt of the analyst to stay as close to the terms of reference as possible. Figure 5.6 (page 57) shows the parallel development of the 'soft' structures and processes model.

With the completion of the 'soft' process and structure diagram we have completed the collection of information necessary for the final composition of the rich picture.

5.2.3 Constructing the rich picture

In one sense this might be thought to be no longer necessary. The foregoing demonstrates that the analyst has got a reasonable grasp of the various areas of the problem context and has sifted out technical from other issues. However, one of the major reasons for producing a rich picture is to visualize the problem situation at a glance. This cannot be achieved if the various elements of the analysis are kept as discrete diagrams.

Rich picturing first requires us to simplify reality. One method for doing this is to set out all the processes and structures, the most important characteristics of the major individuals involved and the terms of reference (see Fig. 5.7 on page 58).

In our case it was useful to separate out the structures and processes with regards to the

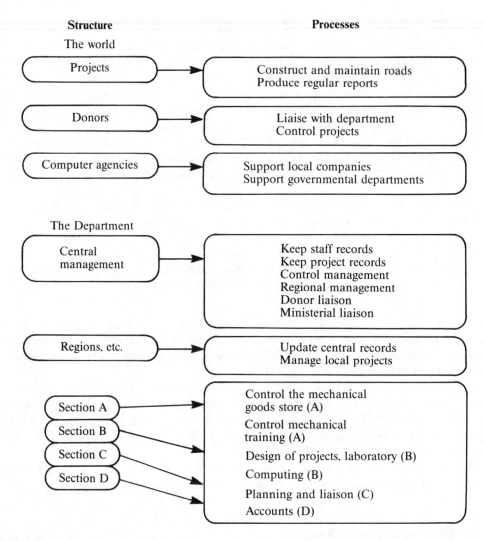

Figure 5.5 Processes and structures—'hard'.

organization on the one hand and the world on the other. Our next task is to set out the major groupings within these two components (see Fig. 5.8, page 59).

We are beginning to bring together all the aspects of the situation into one frame. This in turn provides us with a core concept or '**mindset**' of the problem. One of the most common complaints that practitioners make at this stage is that they cannot draw, or they do not have a clever computer package to produce quality diagrams. Do not worry about this for now. We will give various examples of rich pictures in this chapter, some drawn

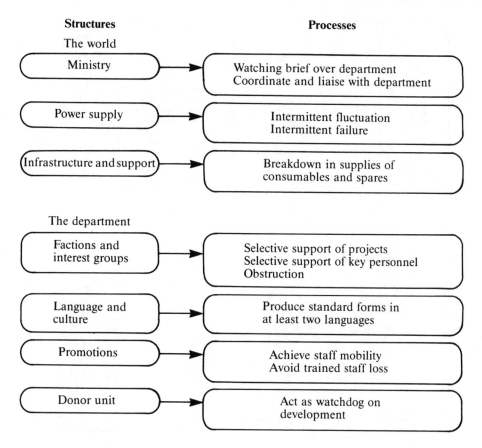

Figure 5.6 Processes and structures—'soft'.

without computers. It is of course useful if a rich picture can be attractive and pleasing to the eye—but of much more importance is the meaning of the content. To make even hand-drawn pictures useful for overhead transparency use it is of value to use a set of symbols that have a clearly defined meaning. In short, to make our final drawings more understandable it is useful to adopt some form of a grammar of symbols. The symbols shown in Fig. 5.9 on page 60 are fairly general; you might think of more for your own situation.

The resulting rich picture for the government department is shown in Fig. 5.10, page 61.

It may be useful to go back to Section 5.2 onwards to be clear in your mind how this picture follows on from the process set out there. All the pictures on the following pages are simplistic and to some extent superficial in their scope, leaving out some of the more contentious details of the previous diagrams. Without the rich picture there is little chance that we could structure into the analysis the type of personal and organizational problems that fall outside the scope of more objective, reductionist forms of analysis.

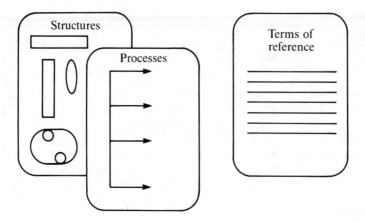

Next we set all of this out in one model.

This is not easy. You may need to take several shots at it.

You could start off by setting out the major themes, e.g.

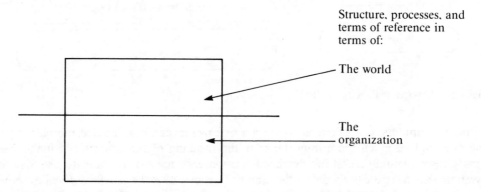

Structure, processes, and terms of reference in terms of:

The world

The organization

Figure 5.7

It should be noted that the rich pictures we develop here and the further analysis that follows are case studies drawn from experiences in various countries. They do not represent any particular department and do not reflect the experiences of any similar institutions.

Remember, the vital ingredient and assumption about the rich picture (and many of the phases that follow) is that they are worked through in collaboration and with the consent of the major stakeholders in the situation, in so far as this is possible, and that the

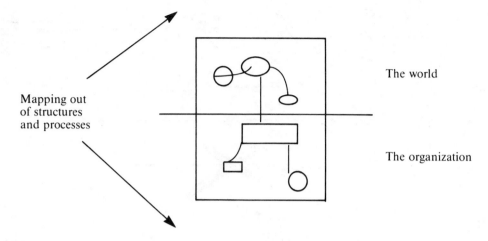

Mapping out
of structures
and processes

The world

The organization

Figure 5.8

information being gathered is not deleterious to the ultimate success of the project. The means by which collaboration and feedback are achieved are:

- Regular reporting,
- weekly/monthly (depending on the scale of the analysis) workshops with stakeholders, and
- regular (daily) discussions and feedback as you work through your thinking.

The examples shown here (Figs 5.10, 5.11, 5.12 and 5.13 on pages 61–65) illustrate what is required in the rich picture. The crosses on lines indicate conflicts of interest or conflicts of some kind between major aspects of the current situation.

For example, in the Department of Roads, the regional offices require rapid briefing on details of roadworks from the projects. At the same time the administration at central office also requires regular reports from the regional offices to ensure that records are kept up to date and the annual ministerial reporting procedure can occur smoothly. Poor communications infrastructure as well as different perceptions of priorities ensures that there is a constant level of friction. The thought bubbles show the main concerns of the major stakeholders involved. You can also see the way in which conflict and competition operate in organizations.

It may take many discussions, workshops, and papers before the picture is agreed. However, this is time well spent because all further analysis work can be more surely directed towards the agreed problem.

Several new points arise from the rich picture:

1. The importance of the power supply issue is re-emphasized.
2. The centralized nature of the department around one key personality is drawn out.

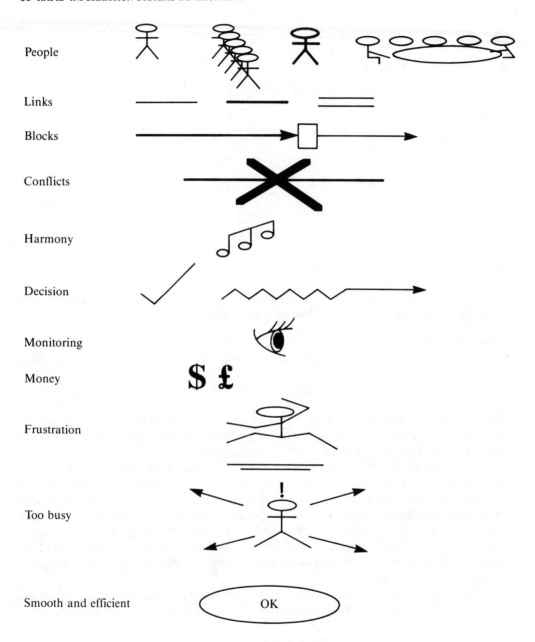

People

Links

Blocks

Conflicts

Harmony

Decision

Monitoring

Money

Frustration

Too busy

Smooth and efficient

Figure 5.9 Easy to use symbols.

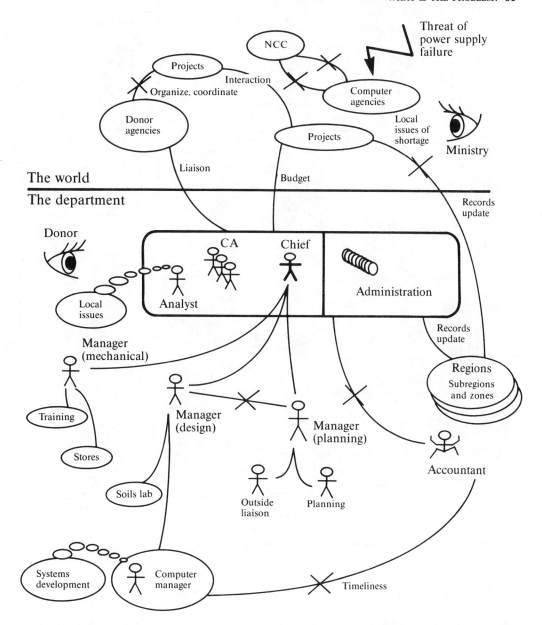

Figure 5.10 The rich picture for the Department of Roads. (NCC, National Computer Centre, CA Computer Advisors.)

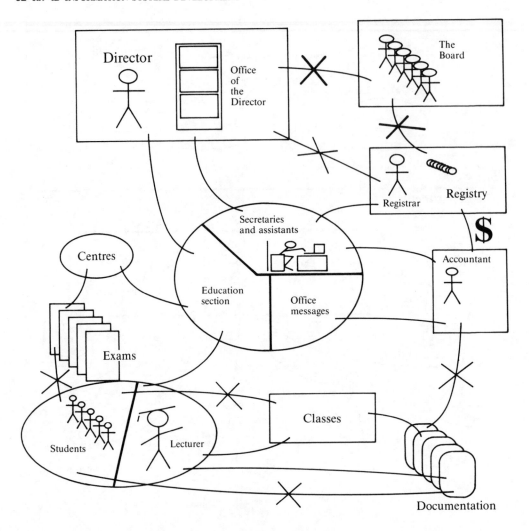

Figure 5.11 Another example: the rich picture of a training institute.

3. The internal conflict between two major sections is emphasized.
4. The peripheral nature of existing computing is expressed.

As a further example of a rich picture, Fig. 5.11 is taken from a different context and is included to provide a further illustration of the variety of forms which the picture can take.

Just to show that this is not the only way in which a rich picture can be displayed, Figs 5.12 and 5.13 show a couple of other examples.

Figure 5.12 shows a training college. The main point made by the picture is the existence

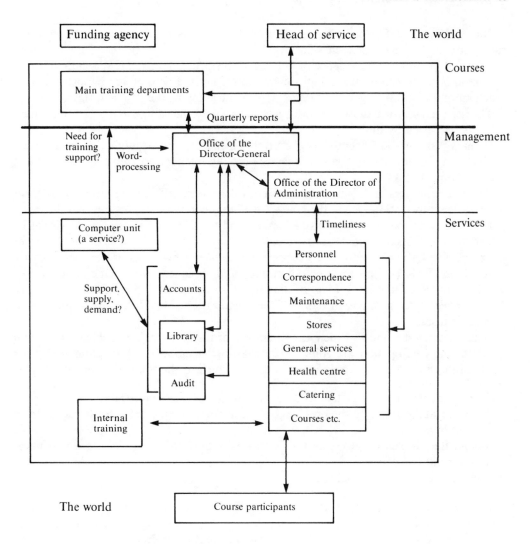

Figure 5.12 An organogram-style rich picture.

of an existing computer unit and the question of its value and relevance. This style of picture can be very helpful for display purposes for senior management. Because the picture is rigidly ordered and very 'neat', this may well have appeal. One word of caution. This type of presentation is often favoured by stakeholders because of its neatness. However, it fails to represent much of the personality-based complexity that a true rich picture can encapsulate. We argue that even a badly drawn traditional rich picture offers the analyst more in terms of depicting problems in the situation being studied.

Our final example (Fig. 5.13) builds on this last point. It is a working copy of a student's rich picture. This may look chaotic and incomprehensible but to the individual who produced it, it contains the essence of his view of his problem context. We show this rich picture in order to demonstrate that presentation is not the item of key importance. It is much more important to get the context and the meaning of the problem right.

The rich picture, when drawn up and agreed, should produce the *primary task* and *issues*. What do we mean by this?

Primary tasks should reflect the most central elements of what we might call the 'problem setting'. Any incoming information system is usually intended to support, develop, and execute primary tasks.

'Issues' are matters of dispute that can have a deleterious affect upon primary tasks. In terms of the information system the issues are often much more important than the tasks.

It is usually not possible to resolve all issues, and for this reason they should always be understood and recognized. Reality really is complex and the analyst should never approach a problem situation with a conceited or inflated view of his or her own capacity. Not all problems can be mapped, discussed, and designed away. Often the analyst will be required to develop a form of amnesia towards certain problems that are either *imponderable* or *too political*, in terms of the organization. This does raise a question of professionalism.

Situations can develop in which large numbers of insurmountable problems arise with issues that, in your opinion, are going to cause lasting impediments to the ultimate systems design. In cases of this type only you can decide which of the following courses to take:

1. Design the information system as best you can within these constraints.
2. Say unpalatable things to the problem owners and set conditions for further work to be carried out.
3. Ignore the problems and create the system as if they did not exist (we would never recommend this course).
4. Refuse to continue the analysis.

Each of these courses has quite serious implications. Only the analyst can make the decision which is most appropriate within his or her own context. Generally speaking we have found that opinion 1 is suitable for 80 per cent of projects.

The bottom line for the rich picture is to provide the analyst with a means to move from *thinking about the problem* to *thinking about what can be done about the problem*.

Amount of time devoted to analysis so far:
Total for this stage (rich picture) = 3 days

Cumulative total to date = 3 days

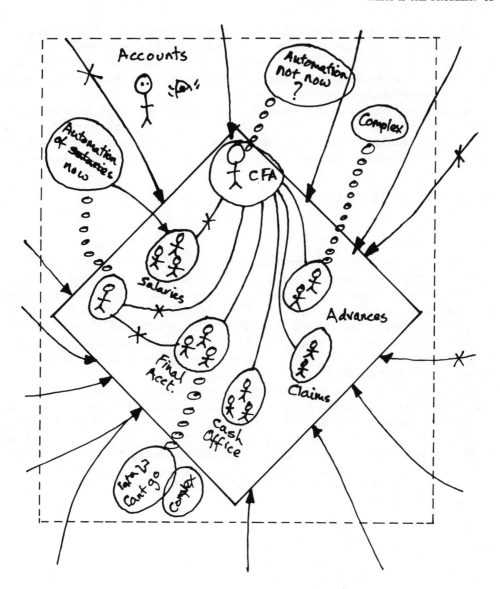

Figure 5.13 Students rich picture.

In most cases we feel that this stage of the analysis can be completed in *less than three days*, though we may need to recognize that rich picturing should not be artificially curtailed. This is one of the most important parts of the systems analysis and sometimes considerably more time is required. If this is so in your case you may need to adopt either path 2 or 3 as set out in Chapter 4.

We can now go on to look at the specific views of the major stakeholders as they are concerned with the new systems definition (What is it supposed to do?). This should result in our tightening up the context of the job we are to do and harmonizing the view of this job between stakeholders. The mechanism we use for this is called the root definition.

5.3 THE ROOT DEFINITION

5.3.1 Introduction

The assumption of the root definition is that the different stakeholders in the system will have different opinions about it. If you were to ask some of the members of a government department questions like: 'What is the main purpose of your department?' you will get different answers, such as: 'To carry out an efficient operation'; 'To keep people employed'; 'To provide a service for the national community'. These are all valid statements of aims, but they may have conflicting implications for the organization and the original terms of reference for the research. Also they are much too vague to help the analyst produce a system that will help the organization in furthering its aims.

It is useful if a point of reference exists whereby the rich picture produced by the analyst and stakeholders in the system can be tested to make sure that the perception of the elements of the contract is being fulfilled. Therefore, at an early stage a careful definition of the required system is essential. Of course this is going to be very general, but in terms of our methodology it contains six ingredients. In terms of the current issue: who is doing what for whom and to what end? In what environment is the new system to be implemented? To whom is the final system going to be answerable? In terms of the HAS these are known respectively as:

- *Client*—the systems beneficiary.
- *Actor*—the individual(s) involved in the system.
- *Transformation*—what the project is intended to achieve.
- *Worldview*[1] (or '*Weltanschauung*')—the assumptions of the project.
- *Owner*—the eventual system owner.
- *Environment*—the situation in which the system will be developed.

This leads to the acronym *CATWOE*.

---

[1] Worldview; in this sense we are referring to fundamental assumptions that effect the proposed information system.

The definition of each of the elements, and the construction of a definition that encapsulates them all, is a matter of negotiation between the stakeholders in the situation, the analyst, and the context of the terms of reference for the project. The forms of communication created during the rich picture stage of the analysis should be very helpful now. Depending upon the time available and the complexity of the situation, you will need to carry out a CATWOE analysis of all major stakeholders. Again, you are the final decision maker in terms of setting out who needs to be questioned.

In our example the definition for the government roads department is shown from the perspective of:

1. The analyst.
2. The donor.
3. The head of department.

From the amalgamation of these with agreement on key items we can come to a consensus view:

4. The department as a whole.

The analyst is in the frame because it is very important to be sure that we are working on the same basic assumptions as the organization. There are cases where the analyst has undertaken systems analysis only to find at the end that the organization was under the impression that the research was being undertaken for very different reasons.

5.3.2 Three examples of CATWOE

The definition of each point of CATWOE can be as drawn out or as brief as you feel necessary. Generally a few words on each item will draw out the main features of each stakeholders views. In our example sometimes only one word is necessary.

The analyst CATWOE

- *Client*: the donor, the department.
- *Actors*: analyst, potential computer staff, actual computer staff.
- *Transformation*: an automated management information system.
- *Worldview*: departmental automation.
- *Owner*: department.
- *Environment*: department and regional offices.

The donor CATWOE

- *Client*: the department/ministry.

- *Actors*: analyst and staff.
- *Transformation*: automated management information system.
- *Worldview*: effective automation for management.
- *Owner*: the department.
- *Environment*: the department.

The head of department CATWOE

- *Client*: the department.
- *Actors*: staff.
- *Transformation*: automation.
- *Worldview*: efficient departmental operations.
- *Owner*: head of department.
- *Environment*: department.

These three views were supplied during interview, as were most of the details of the rich picture. They offer us a fair degree of agreement within the problem context. It is the analyst's job to assess the degree of differences between root definitions and to harmonize an overall view that all stakeholders can agree to. This will mean that differences in interpretation will not occur (or are less reasonably likely to occur!) later on. In some cases the root definitions can be seen as fixing together to form a cone focusing on the problem situation at the root of the exercise, as shown in Fig. 5.14.

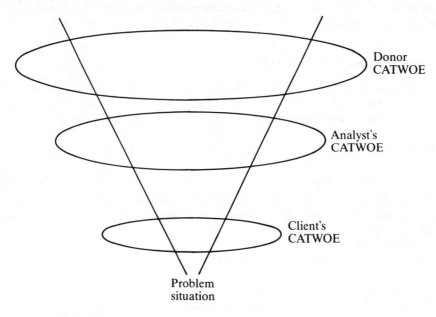

Figure 5.14 Overlapping root definitions of the problem.

In this case the three levels of root definition can be seen as being focused on one agreed problem situation. The graphic presentation seen in Fig. 5.14 indicates that differences in CATWOE relate more to the position of each party (remote or close to the problem situation) rather than marked differences in opinion about the problem. From this we can produce a consensus view.

The department as a whole (a consensus view)

- *Client*: the government.
- *Actors*: analyst and staff.
- *Transformation*: to set up an automated management information system.
- *Worldview*: to automate the department.
- *Owner*: the donor (remote)/department (immediate).
- *Environment*: the department.

Of course, this consensus view has to be agreed by all major stakeholders involved. Agreement may require a certain amount of flexibility by all parties.

The establishment of an agreed root definition takes us to the point where within the context of the situation as set out in the rich picture, and the agreed perspective of the root definition, we can begin to design our new, improved systems outline. This outline is set out in the conceptual model.

Amount of time devoted to analysis so far:
Total for this stage (root definition) = 1 day

Cumulative total to date = 4 days

In most cases the root definition can be arrived at in as little as one to as much as five days (depending on the complexity of integration). There can be exceptions to this rule. Figure 5.15 overleaf depicts a very different view of a root definition.

In this case there is very little agreement among the various stakeholders as to what should be done. The various views could be brought together into a rough consensus as shown in Fig. 5.16 on page 71. This example would appear to be doomed to failure. If this were so, then the root definition has served us well, showing up major structural weaknesses in the new information system plan.

5.4 THE NEW CONCEPTUAL SYSTEM

5.4.1 Introduction

The rich picture is intended to be 'rich' in terms of people, processes, ideas, conflicts, etc.

Director of research organization (RO)

Client: self and RO
Actor: key staff in RO
Transformation: composite data translated into information, quickly
Worldview: 'we have the data, not the information'
Owner: self
Environment: RO

Head of research department in RO

Client: 'him' (the Director) — 'one of his whims'
Actor: 'all to likely to be me!'
Transformation: 'create more bloody work when we cannot cope now?'
Worldview: 'I have data and information but no time'
Owner: 'him' (the Director)
Environment: 'as far away as is possible'

Funding agency

Client: RO
Actor: key staff in RO monitored by funding agency
Transformation: improve effectiveness to forecast and act against drought
Worldview: we have the technology
Owner: RO
Environment: RO

Analyst

Client: 'him' (the Director), the funding agency, and all departments
Actor: initially me — to be all departments focused on IT unit
Transformation: create more work and improve effectiveness
Worldview: questionable technology — a 'test case'
Owner: the Director
Environment: the RO

Figure 5.15 An example of incompatible CATWOE: geographic information systems (GIS).

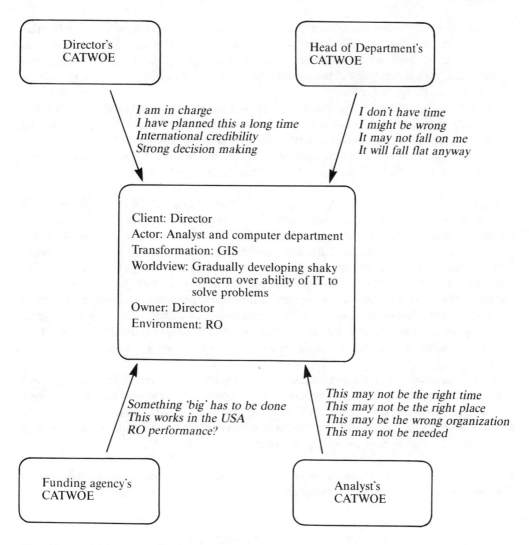

Figure 5.16 Resolving the problem?

Once you have a feel for the problem context we can begin the process of drawing out the aspects on which we now know we have to concentrate. The intention of this phase is to build a model of the system that we recognise as being a reasonable basis for the new information system.

There are two items that we must be aware of:

1. The desire to take elements of our methodology out of context. It is not our purpose here

to specify exactly what the system must do, who will do it, and how long it will take. This process might more usefully be thought of as arising in the second and third phases of the methodology, which deal with information modelling. At this stage we are only concerned with separating out the main component activities of the system and to show how they relate to each other.

2. Beware of the tendency to assume that we are modelling reality. All models are symbols of reality and represent the assumptions of individuals and groups. In this case the model will *represent* the shared perception of the activities of the ultimate information system as focused and presented by the analyst in collaboration with the major stakeholders. This is quite a different thing from saying that we are modelling reality.

Back to the modelling exercise. Ideally, and according to all the textbooks, we should remove personalities, though not their roles, from the picture, because we do not want to create a system around particular personalities. Frankly, this is not always possible. Many organizations are so designed around key personnel that to design any incoming system without taking them into account would be a nonsense (you will probably have identified whether this is so in your case during the phase of the rich picture). At best we can say, therefore, that at present we should, insofar as is possible and useful, sideline personalities from our systems design. In phase three of the methodology we will be looking to integrate the computerized information system into the lives of the people who will be using it.

In the creation of conceptual models we should recognize two key issues:

- It is not the job of the model to align existing subdepartments/units with tasks. At present we want to set out the incoming system tasks irrespective of units and sections. The new system may well require the substantial reworking of such groupings.
- The model has to comply with the results of our *root definition* and our *original terms of reference*. It is quite easy to get carried away at this point!

If you are having trouble assessing the dynamics of the new systems design we recommend that you carry out a two-stage conceptual modelling phase.

1. Produce a model that depicts the existing system.
2. Produce the model that demonstrates the improved situation.

Note that this can only be carried out successfully if you are given the necessary time to do the work (you will know your own constraints) but if this is not possible go straight on to the second activity.

The development of our conceptual model takes various stages.

1. Reassess the consensus root definition to form an impression of the type of system that will be necessary to carry out the transformation generally agreed to.
2. Put together a list of verbs that describe the most fundamental activities of the defined system, e.g. record, liaise, purchase, report, inform.

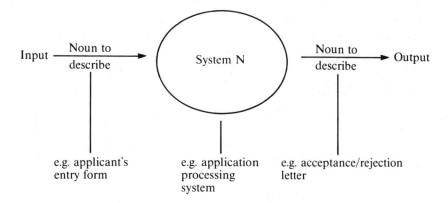

Figure 5.17

3. Thinking in terms of a simple system (input, process, and output), structure what each component part of our system will have to do, how it has to do it, and how activity will be monitored. Use nouns to describe the input and output where appropriate (see Fig. 5.17).

This is a simple idea of a system. Usually systems are developed with regards to the emergence of new properties as items are combined and the hierarchy by which these items are related. For our purposes in setting up information systems practically and rapidly we simplify the issue. For a fuller exposition of the theory see either of Checkland's books (1983, 1990).

4. Structure similar activities into groups (e.g. day-to-day accounts, long-term budgets, short-term budgets could be grouped in a financial system) (see Fig. 5.18 overleaf).
5. Use lines or lines with arrows to join the activities/systems together. The arrows symbolize information or energy or material or some other form of dependency. It is quite useful to use the arrows to represent the main flows of information between systems. The information output from one system is usually the information input for another (see Fig. 5.19 overleaf).
6. Verify the model with the users of the existing system. This is very important. The relationships and major inputs and outputs need to be agreed with all major stakeholders in the system.

Figure 5.20 (see page 75) shows a conceptual model of the government roads department. Note how it has been derived from the corresponding rich picture in Fig. 5.10. The boundaries show the various subsystems within the overall organizational system.

In a particularly complicated or large organization (i.e. in most cases where an analyst is being used in the first place) there is the need to produce various levels of model. Figure

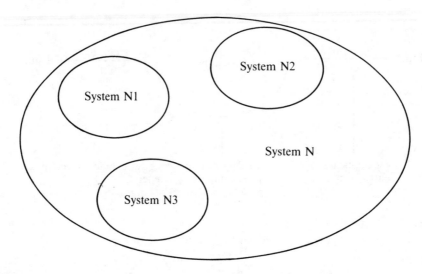

Figure 5.18

5.21 is a level 2 model showing more of the details involved with the management and administrative system as shown within the level 1 model.

The level 2 model shows quite clearly the central role of the head of department office and the immediate subsystems which serve that office. Each of these would ultimately need to be further developed in a similar manner to give us the actual workings of each unit. The

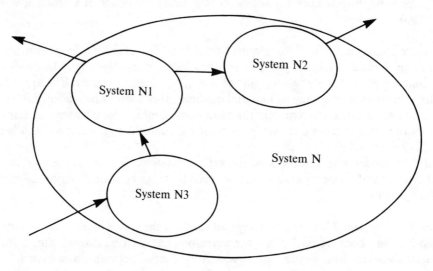

Figure 5.19

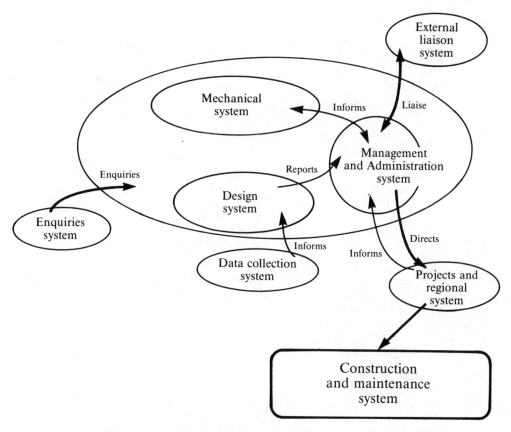

Figure 5.20 Level 1 conceptual model arising from the rich picture of the Department of Roads. Comprising management, administration, accounting, planning, foreign interaction.

conceptual model gives us a pad from which we can launch our detailed information-modelling exercise.

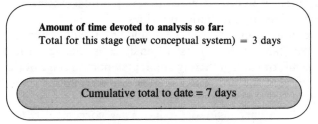

Amount of time devoted to analysis so far:
Total for this stage (new conceptual system) = 3 days

Cumulative total to date = 7 days

The time schedule for the exercise can vary but *if no more is required in the early stages of a project than an understanding of the major activities taking place, a period of between two and five days should be sufficient.*

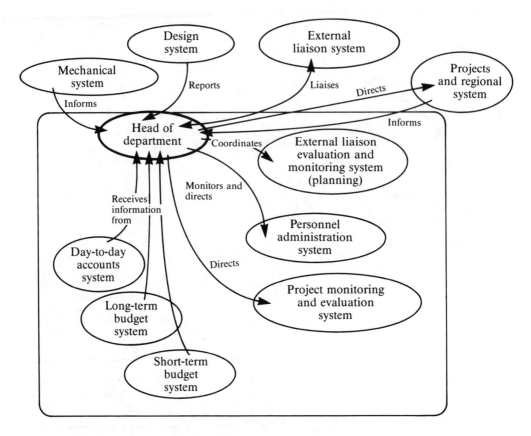

Figure 5.21 Level 2 conceptual model: the management and administrative system.

Before we can do this, however, we must identify the items arising within the model that have priority in the development of our terms of reference. It is unlikely that we will initially be able to carry out all the exercises that would be required to set up a total system. For this process to be effective we need to identify our initial priorities.

Figure 5.22 shows where we have come from and what we should be left with at the end of the first stage of our analysis. We can see the process of analysis as coming from a situation of ignorance. Through the information gathering and representation stage of the rich picture we build up an idea of what the problems are. The root definition further focuses us on an agreed perception of the major components of the work in hand, most importantly the transformation. Finally (at this stage) we produce a new conceptual system of the activity we wish to develop.

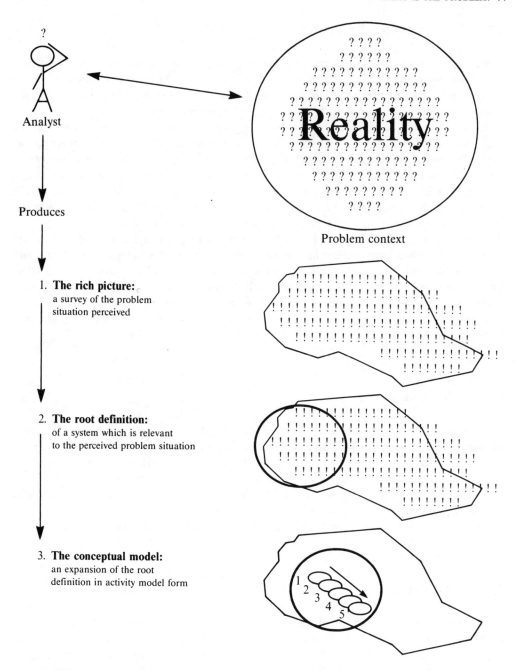

Figure 5.22 Selecting priorities—the process so far.

5.5 FINAL CONSIDERATIONS

The terms of reference will make you very selective in your initial problem area identification. Thus, in terms of our CATWOE, the first person who has a say in this may not be the problem owner in the immediate sense but the more remote owner (if there is one in the case you are dealing with) who is funding the overall operation.

The first job will be to define the boundary between the activities to be included in the analysis and design and those that are pertinent, may well be candidates for further phases within further projects, but are outside the short-term priorities of the terms of reference. The exact position of the boundary must be a matter for discussion between the analyst and the stakeholders. Often, however, the analyst will be asked to advise on areas of the organization's operations where a new information system could produce the greatest benefits. The area that the analyst selects may be the only one to be tackled, or it may be the first part of a phased study of information processing throughout the organization. However, we should keep in mind before we go into technical detail about what should and should not be computerized that often political and other interests will preclude the analyst from dealing with the real area of concern at all. All examples of management interference with the analyst's area of study, specifying certain areas out of bounds, demarcating 'priorities', etc., can be seen as political acts, but as we have already noted these often have to be accepted by the analyst as facts of life.

5.6 CONCLUSIONS

By the end of this stage of the analysis you should have moved from a position of seeking to identify the problem situation to having an organizationally shared view of the potential model of a solution. The rich picture gave the overview, the root definition defined the key issues and identified the primary task, the conceptual model has outlined the next step. We can now enter the 'hard' phase of our methodology and look at the process of information modelling. In this stage we will seek to indicate the major components of the proposed system in a manner that can be readily transposed into a working system.

5.7 TUTORIAL

This is the first tutorial exercise. We are assuming that you have an existing organizational problem and would benefit from an exercise, set out in sequence, of how to go about the process of analysis and design. Similarly, if this book is being used as a class text the following tutorial can be used for the core of an assessed exercise. A model approach to the problem is set out in Appendix 3.

Exercise 1 The human activity system phase

Sequence of action You are a consultant reviewing the capability of a construction company in a developing country to make effective use of automation. Here are some details.

Personnel

Managing director—M. R. Smith.
Director—D. F. Jones
Director—A. F. Smith.

Departmental head (A, policy)—G. T. Brown.
Departmental head (B, works)—G. V. Smith.
Departmental head (C, planning)—B. T. White.

The department employs 89 people in the head office and 656 people outside the head office.

Tasks

1. Preparing quotes and outline contracts (buildings, roads, emergency repair to river banks and sea defences, rail).
2. Dealing with subcontractors.
3. Drafting contracts.
4. Project supervision.
5. Deadline enforcement.

National characteristics

1. Poor regularity of power supply.
2. Poor road and rail transport.
3. Difficult topography and seasonal heavy rains, which cause further transport difficulties.
4. Poor internal telephone service. Supplemented by radio communication.
5. Scarce access to hard currency.

Your brief is to look for likely departments within the company requiring effective management information system tools and which have a high probability of being able to maintain a new system.

You have been able to glean other useful information.

- Company turnover has been static for the last three years; there has been a linked reduction in customer demand.
- The accounting section has been seriously undermanned for three years and has suffered from considerable labour loss (particularly younger staff).

- The senior accountant does not sit on several senior committees.
- There is a small computer unit using very old machines and turning out very poor payroll and costing information.
- Morale among senior staff can be seen as being fairly poor.
- One family has members in several senior positions. This family link tends to be the information/operational spine of the company.
- Junior staff are generally well trained but frustrated by poor promotion chances.
- The outlook is good, contracts being negotiated would indicate that workload will increase at 7 per cent per annum for several years.

The three major departments within the company—planning, policy, and works—show a certain amount of internal friction. Loyalty to family appears to be a bone of contention.

The planning department deals with contractual details and some works design.

The policy department exists to lead discussion with major customers (government departments, private companies) and contractors and to set outline policy statements.

The works department is the business end. It employs 75 per cent of staff, carries out and/or supervises construction and maintenance. Works is directed by one of the Smith family.

Works employees operate all regional offices.

Your project donor/financier is an international bank. They are looking for an area to invest $300 000 initially in management information system activity.

Given this background, produce the rich picture for the organization. Prepare the rich picture as a brief for yourself and possibly as the basis of a seminar to brief the donor.

Some hints You have three key areas—the world, the regional offices, and the department. Obviously there is conflict in the department and the prominence of the Smith family cannot be overlooked. As this is a document for the donor you can be quite frank in your views of the situation.

Do not attempt to appear to be all-knowing. There will be a lot of details that you will not have. For example, how does the organization's administration fit into all this? What role does the existing computer unit have and how well trained are its staff?

Be sure to make a list of items that you will require more information about.

Exercise 2 On root definitions and conceptual models

Working from the rich picture go on to prepare root definitions for:

1. The analyst (you).
2. Hypothetical—for the donor.
3. Hypothetical—for the managing director.

Include all your own doubts and problems with the job in your own CATWOE. Remember, how do the terms of reference fit with what you have found in the rich picture? Do the regions need help? Are they getting it? Should other individuals and agencies be questioned?

Presumably if (for example) the analyst finds that the regions could do with computer support in the line functions of the organization, whereas the client and bank feel that the core concern is central office MIS detail, this will show up in your consensus CATWOE.

Outline the top-level conceptual model of the initial MIS you would set about designing. Pay particular attention to:

1. The limits of the initial MIS.
2. The products of the initial MIS.
3. The dangers for the initial MIS.

Your existing work will probably have indicated that the organization contains numerous subsystems (e.g. management, strategy, policy, planning, and works). Your MIS will need to focus on these types of subsystems. How will you deal with the 'issue' of the regions?

A model answer to the exercise is given in Appendix 3, Sections 1–3.

FURTHER READING

Anthill, L. and Wood-Harper, A. T. (1985) *Systems Analysis*, Made Simple Computer Books, Heinemann, London.

Avison, D. E. and Wood-Harper, A. T. (1990) *Multiview: an exploration in Systems Development*, Blackwell Scientific, Oxford.

Avison, D. E. and Wood-Harper, A. T. (1991) Information systems development research: an exploration of ideas in practice. *Computer Journal*, **34**, No.2.

Carter, R., Martin, J., Mayblin, B. and Munday, M. (1988) *Systems, Management and Change: a graphic guide*, Paul Chapman Publishing Ltd., London.

Checkland, P. B. (1983) *Systems Thinking and Systems Practice*, Wiley, Chichester.

Checkland, P. B. (1984) Systems thinking in management: the development of soft systems methodology and its implications for social science. In R. Ulrich and G. J. Probst (eds), *Self-organisation and Management of Social Systems*, Springer-Verlag, Berlin.

Checkland, P. B. (1985) From optimism to learning: a development of systems thinking for the 1990s. *Journal of the Operational Research Society*, **36**, no.9, pp. 757–767.

Checkland, P. B. (1988) Information systems and systems thinking: time to unite? *International Journal of Information Management*, no.8, pp.239–248.

Checkland, P. B. and Scholes, J. (1990) *Soft Systems Methodology in Action*, Wiley, Chichester.

Wood-Harper, A. T. (1990) Comparison of information systems approaches: an action-research, Multiview perspective. Ph.D. thesis, University of East Anglia, Norwich.

6

INFORMATION MODELLING: MAKING A WORKABLE SYSTEM

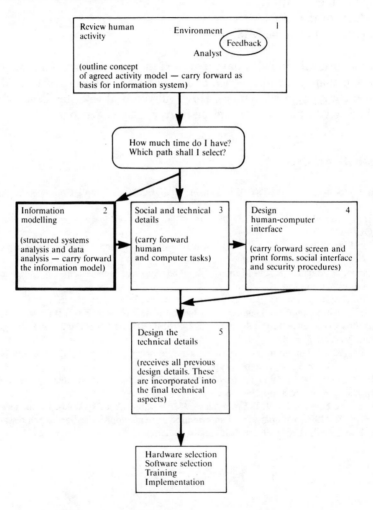

Figure 6.1 Rapid planning methodology.

Keywords entity model, functional decomposition, events, entity/function matrix, data flow diagrams, attributes.

Summary The conceptual model leads on to a harder systems analysis and design, i.e. information modelling. During this phase the subject of the problem is broken down in terms of entities, functions, attributes, and events. The ground is prepared for the proposed information system in that entities correspond to things that we wish to keep information about. Functions are the jobs that these entities are involved with. Attributes are the qualities (or fields) that compose the entity, and events are the triggers that cause functions to arise.

6.1 INTRODUCTION TO INFORMATION MODELLING

The second, 'harder', stage of our methodology is to structure our conceptual model of the new information system in such a way as to produce a workable information system outline for the major stakeholders. Once the analyst and the stakeholders have reached agreement (even if this is quite tentative) on the overall picture of the situation and on the root definition of the system to be designed, then information modelling can begin. This stage can be very long winded. To achieve a high measure of accuracy in terms of information modelling the task can take a considerable amount of time to work itself out in great detail. This is not the purpose of this book. Texts that offer a high level of academic accuracy are indicated in the further reading section at the end of the chapter. In terms of the present task *we do not want to spend our time on lengthy academic review; we need to think about what our information system is actually going to do and attempt to produce an outline system that is practical and workable.*

The conceptual model on the following page (Fig. 6.2) gives us the broad brush strokes to work from. We now need to identify (in liaison with the stakeholders of the system):

- What do we wish to keep records about (entities)?
- Of what are these entities composed (attributes)?
- What functions are carried out by the entities?
- What are the triggers or events which fire the functions?

Of course, at any one stage in the analysis and design process it is impossible not to think of the manner in which the current stage will affect those that follow. This is a very useful feature but one that analysts and designers in the past have undervalued. It can be argued that at each stage of the analysis it is best to attempt to banish all past and future analysis from your mind. This ensures that analysts do not attempt to take elements of the analysis out of context, e.g. to specify the hardware and software for a system before having carried out the information-modelling stage. The problem with not thinking about other stages is that the links between stages are ignored and we end up with a five-stage methodology, with five stages, none of which interrelate with the others. Therefore our task at this stage, and

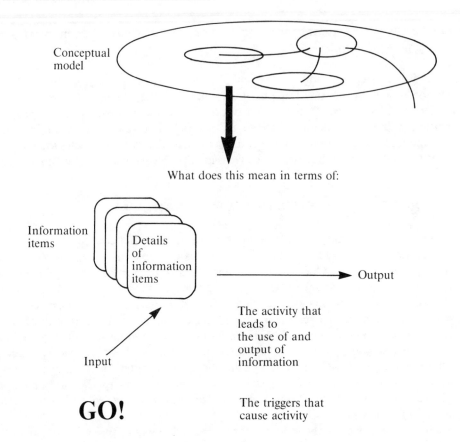

Conceptual model

What does this mean in terms of:

Information items

Details of information items

Output

Input

The activity that leads to the use of and output of information

GO!

The triggers that cause activity

Figure 6.2

at every other stage, is to keep a clear idea of the ideas that led to the current analysis, keep in mind the main needs of the stage to come, and concentrate chiefly on the work in hand (see Fig. 6.3 opposite).

To be aware of how our current work fits into our overall analysis is useful, but to make any decisions concerning hardware, software, and training would be to reduce the value of the entire analysis exercise and would make a nonsense of the process. We will eventually want to make decisions in terms of *implementation,* but this is not the time. There is great value in having a nominal idea as to what combinations might produce the system we are designing, but these ideas should be kept to ourselves until the final stages. The reason for this is quite simple. Any joint discussion of hardware and software with stakeholders can raise false expectations, unnecessarily bring pressure to bear on yourself to deliver, and most importantly, *rule out the possibility of making changes to your planned system. If stakeholders perceive the system that they have so far agreed to is to be changed they may consider*

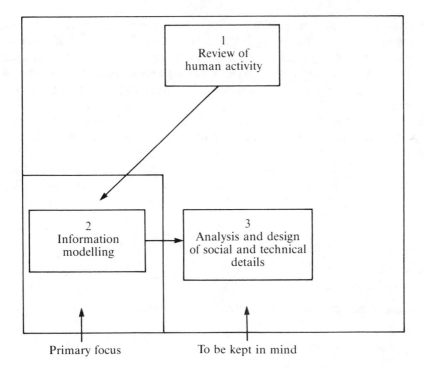

Figure 6.3 Think back and think forward.

this to be regressive. This in turn can cause problems in the relationship between the analyst and the stakeholders.

With these thoughts in mind we will now look at the major features of the second stage of our analysis.

6.2 ENTITIES, ATTRIBUTES, FUNCTIONS, AND EVENTS

There are four features to which we are going to try to reduce our proposed information system. In order for our analysis to be accurate we need to spend a little time defining them.

1. An entity—defined as a thing that records are kept about. The definition is intentionally vague in its meaning. The need for a degree of flexibility is essential because the entities in a system can range from the major individuals who are working within the organization (e.g. senior managers, chief accountant) to strategic information sources (e.g. staff records, sales records, payroll, land use data).

2. Attributes, i.e. the attributes of the entities. For example, if the entity is a land use planning system, the attributes might be rainfall, climate, percentage of land use arable, percentage of land use pasture, percentage of land use urban. On the other hand, if the entity is a management information system dealing with company performance, some of the attributes might be gross sales for years x to y, gross profit, net profit, number of staff employed, staff salaries.
3. Functions are actions that take place concerning entities. Therefore some of the functions related to a small computer maintenance business might include update customer ledger, update supplier ledger, keep inventory of stock, register sales, register bad debts, register sales staff mileage. Functions set out in this form are fairly chaotic. To understand the functions that make up even a basic operation we make use of a hierarchy tree. For example, the major function:

Update customer file

would contain such sub-functions as:

Receive sales data
Receive bad debt data

The first of these items might then contain sub-subfunctions like:

Add new customer data
Edit old customer data
Delete archive customer data

Of course, this is a very simple example, and quite often you may need to go into quite a lot more detail. The breakdown of the functions or functional decomposition, will work its way out to a tree structure (see Fig. 6.4).
4. Events are trigger activities that make functions occur. For example, in a training organization a potential participant for a short course wanting to enrol is an event that triggers the function 'process application'. This may in turn trigger off other functions —'check vacancies' or 'assess sponsor'.

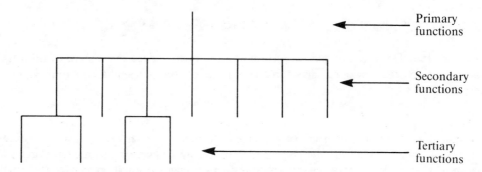

Figure 6.4 Tree structure.

If we try to put the entire scheme together in one particular case it might appear as follows:

ENTITY   management information system

has

ATTRIBUTES   staff records
                        sales records
                        purchase records

carries out

FUNCTIONS   annual staff appraisal
                        strategic planning

at particular

EVENTS   end of financial year
                        any strategic event
                        random events

This is, of course, a simplistic example but does show the way the entire system fits together. We will now go on to look at each element in greater detail.

6.3 ENTITY MODELS

Entity models are usually the primary components to look at. If we do not know what we want to keep records about, then we have not got a conceptual view to work from.

It is possible to start off the analysis with a review of functions or entities. We feel it makes sense to begin with the entity (the noun so to speak) rather than the function (the verb). Also, for the purposes of the continuity of this multiperspective methodology, we feel that the clearest correlation between the HAS and information modelling is from conceptual model to entity model.

What are the major items about which we wish to store information? It should be remembered that all entities have functions and attributes and we will eventually want to link all these items together. The process of arriving at a definitive mapping of entities can be argued to be a slightly theoretical exercise as there is room for judgement in the selection. Two things are important and have been pointed to many times by other authors: getting a complete picture and at the same time not flooding the analysis with information. We will discuss this in greater depth shortly. The process of producing a map of entities and groups of entities (known as entity types) can be seen in Fig. 6.5 overleaf.

This is far from being a straightforward process. You may need to simplify your analysis down to one or two basic entities. There are a few basic rules that need to be understood concerning relationships and entity models. (See page 88, Fig. 6.6.)

Procedure

1. List all entities for the new information system

2. Group related entities into blocks (entity types)

3. For each block:
   (a) draw the entities as boxes
   (b) link the entities together
   (c) show on the link line the nature of the relationship

4. Put all the blocks together

5. Link the blocks and indicate the nature of the relationships

Figure 6.5 Entity map procedure.

Earlier on we mentioned being complete in terms of our entities while at the same time avoiding flooding. By completeness we mean that no major thing about which you wish to keep information is missing. Flooding refers to the potential complexity of the final model if we were to map out every single entity that comes to mind. At this stage we want to map out major entities only. In fact the process of reducing complexity down to the key

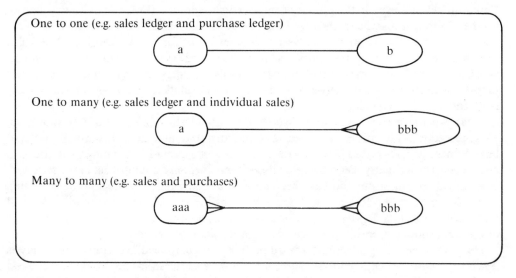

Figure 6.6 The various types of relationship in an entity model.

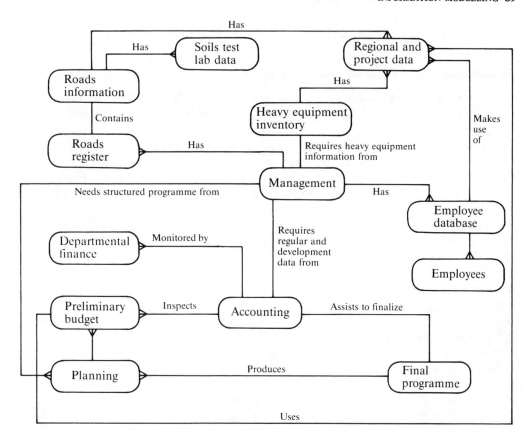

Figure 6.7 The entity model.

components of study is an important lesson to learn. It will always be possible to increase complexity and even have levels of entity model later on. In Fig. 6.7 we have linked all the entities by means of lines and notes defining the relationship. Figure 6.6 indicates what these relationships mean. If the line has a multiple line ending (crow's feet) it shows a one-to-many relationship (e.g. the employee database contains data on many employees). There are three types of relationship between entities:

- One to one (e.g. employee and salary).
- One to many, many to one (e.g. personnel database and employees).
- Many to many (e.g. projects and personnel).

Perhaps the easiest way to understand this is to look at an example. Figure 6.7 is the entity model for the government roads department.

Sometimes the analyst may find it helpful to make a note of the direction of the entity relationship, e.g.

department finance monitored by accounting

The reverse (accounting monitored by department finance) does not seem to be too silly, so confusion might arise without this type of clarification. The name for this type of explanatory note is 'anchor point notation'.

The major insight that we can give concerning your entity model construction is that it may take many attempts to get the model clearly defined. Lines should cross as little as possible, and for most purposes you should be able to reduce the situation to about 20 entities. If there are more than this you are probably dealing with a very complex problem and unless you have existing analysis and design skills we recommend that you try to begin by reducing the scope of your analysis to a subset of your original terms of reference. In our example here we have reduced our analysis to four key areas as derived from our conceptual model:

- Programme planning and finance.
- Personnel records.
- Roads register.
- Equipment inventory.

When you are satisfied that you have:

1. identified and listed your major entities, and
2. modelled them in a similar manner to that given above

you are ready to go on to the next stage. It is quite important to realize that entities can change into other entities, e.g. for the training institution an applicant could become a student. In the example we have above the accounting entity transforms the preliminary budget into the final programme entity.

Another example of an entity map is shown in Fig. 6.8. This relates to a training college and shows the map for a project focusing on staff student records.

It is not the purpose of this book to go into considerable detail on these issues. Entity modelling (and all the following stages of this phase) can be developed in great depth. If you feel that you require greater detail we refer you to the further reading for data analysis. Our main theme is not theoretical purity but to make use of existing analysis and design tools and produce practical and reproducible analysis and design products.

At this stage it is important that you keep a note of the potential size of entities—i.e. will they contain hundreds, thousands, or more information or data elements? This information will be very important for the next stage of the analysis.

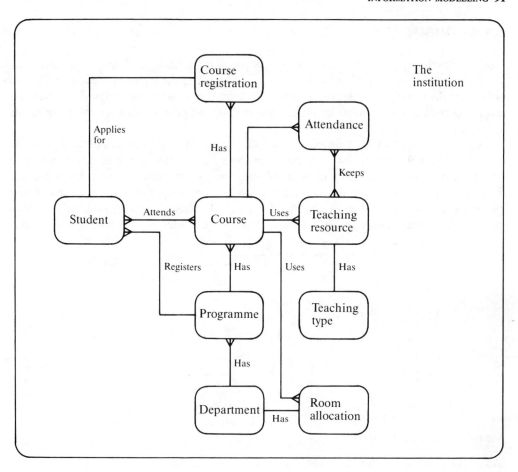

Figure 6.8 Entity maps of a training college.

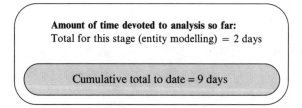

Amount of time devoted to analysis so far:
Total for this stage (entity modelling) = 2 days

Cumulative total to date = 9 days

Working on the estimates of time we set out in the first stage of our approach, we suggest that two or three days should complete this phase of stage 2.

For now we are assuming that you are ready to go on to mapping the major attributes as related to the entities.

6.4 ATTRIBUTES

For the development of the information system it is useful if we begin to identify key attributes of entities. Thinking back to our introduction to the chapter, the reason for this is because the attributes we set out here should form the basis for the fields of our eventual data-base for the management information system (in the primary case we are using in this example).

One way of carrying out this exercise would be to create an entity/attribute matrix, as shown in Fig. 6.9. That is, for each entity on the x axis we list all the attributes on the y axis to make a complete listing of all attributes. One reason for *not* doing this is because if we have quite a lot of attributes related to any one particular entity, the matrix would become rather ungainly. Ten or eleven entities would produce possibly hundreds of attributes. It is probably easier just to set out each set of attributes against each specific entity.

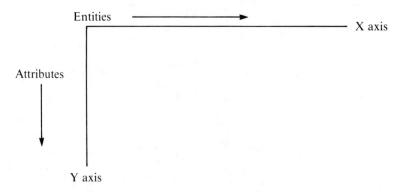

Figure 6.9 Entity attribute matrix.

An attribute listing is shown below.

Attributes for the roads register entity

Name of road.
Date of construction.
Personnel involved—engineers, overseers.
Duration.
Cost.
Benefit as projected in original report.
Source of finance—external, internal.
Total quantities and costs of:
    cutting
    filling
    gravelling

culverts
bridges
etc.

Obviously we could go into much greater detail. The amount of detail you need is again your decision. It is not intended that at this stage you should be thinking of the actual database structures that might be required to accommodate the datasets that you begin to generate in outline. A rather different approach is seen on page 94 in Fig. 6.10, the attribute listing for the training college entity map.

This example of attribute mapping focuses on setting out the analysis as an aid to computerization. Each entity can be clearly seen as a record and a group of entities is a file. Each attribute is a field in an entity. Sometimes this type of approach is appropriate. For example, if the analyst knows that he or she is required to design a computer-based system and is familiar with database design, there is little point in not setting out the attributes in this manner. However, if a manual system is required, or if there is some debate as to what the analysis is actually telling us, then it is better not to be thinking of the final system at this stage. Again we cannot overstress the need to resist the impulse to attempt to plan the whole system before carrying out the entire analysis.

As with the entity phase of this stage, keep a note of the likely size of the attributes for the entity—i.e. how many attributes are there, how many records will be kept in the system?

> **Amount of time devoted to analysis so far:**
> Total for this stage (attribute listing) = 1 day
>
> > Cumulative total to date = 10 days

One day should see this phase completed.

6.5 FUNCTIONAL DECOMPOSITION

This may sound rather a mouthful, but as with so many analysis and design terms the reality is quite simple. **Decomposition** as used here refers to an hierarchy of tasks broken down into their component and even subcomponent parts. This is used to show the major functions and the way in which these consist of other simpler functions. A simple example would be to demonstrate digging a hole in a road (see Fig. 6.11 page 95). This is a trivial example but demonstrates the way in which decomposition works.

The breakdown of the whole into its parts is known as top-down decomposition. In Figs 6.12 and 6.13, on pages 95 and 96, we demonstrate the decomposition of functions at two levels for the department of roads.

Student	Course	Teaching resource	Course registration
Attributes	Attributes	Attributes	Attributes
Registration number	Course title	Name	Registration number
Name	Abbreviation	Address	Course title
Home address	Contact hours	Full time/part time	Year taken
Term address	Department	Contact hours	Course mark
Department	Potential numbers	Department	Exam mark
Full time/part time		Expertise	
Contact hours		Grade	
Fees paid		Title	
Programme		Actual contact hours	
Result			

Teaching type	Room allocation	Attendance
Attributes	Attributes	Attributes
Grade	Room number	Course title
Hours	Department	Year
Salary	Seating capacity	Weekly attendance
	Actual usage	Term attendance

Figure 6.10 Attributes.

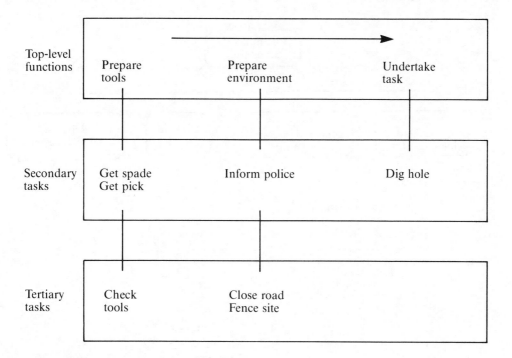

Figure 6.11 Simple functional decomposition.

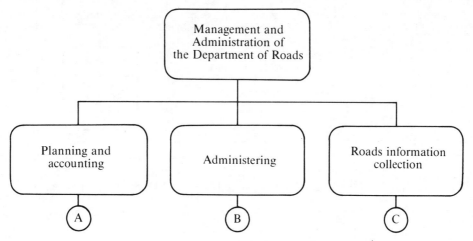

Note: each box is labelled as a function, rather than an entity
(e.g. 'Administering' not 'Administration').

Figure 6.12 Top level function chart—the department of roads.

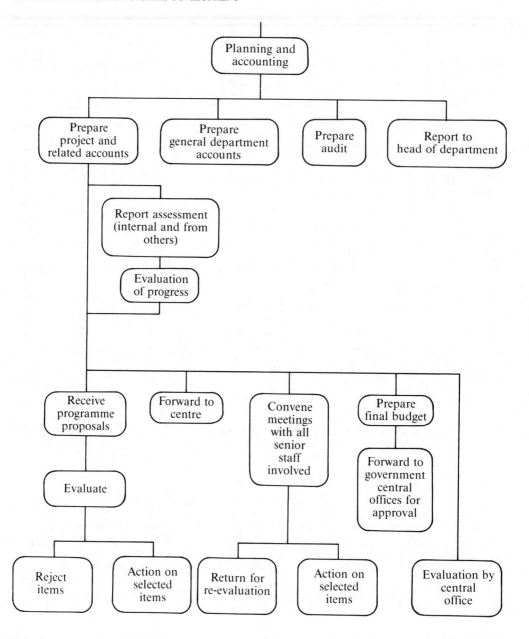

Figure 6.13 Level 2 function chart—planning and accounts.

It is interesting to notice how there are three major functions—planning and accounting, administration, and the roads information collection. This does not mean that these are the only functions. It means that these relate most closely to the job we have been set in our terms of reference, the greatest area of need as shown in the rich picture, the consensus view arrived at in the root definition and the new systems outline as given in the conceptual model.

It is hardly surprising to realize that the major areas proposed functionally for a management information system in a government department are administration and major inventories. Also the planning and accounts areas are fairly obvious contenders for new information systems design, as invariably our priorities will be initially focused on repetitious and well-structured tasks.

The level 2 chart develops the decomposition of the planning and accounting function. The point to note on this chart is that there are two paths to 'receive programme proposals'. Ideally this should always occur following the assessment and evaluation stages. However, there are times when this is not possible and an informal short cut is taken. This leads us into the area of the discretion of the analyst and professionalism again. In some cases it is the responsibility of the analyst to remove unwieldy pieces of activity and in others to conform the new design to some tried and tested (if 'informal') activities. The way in which the individual analyst deals with this type of 'informal' information processing reality entirely depends upon the specific situation in which he or she is working. Sometimes the problem owner may not wish to have this pointed out but will still require you to work round it. It is another example of the analyst having to sometimes make use of a subjective evaluation of an 'economy of truth' in terms of his or her actual reporting.

In our other example of a training college, functions were mapped out much more closely with the final computer system in mind (see Fig. 6.14 overleaf). This does require a little explanation. The example is very specifically related to generating information products (**performance indicators** or PIs) for a training college. Much of the detail is abbreviated (FTL = full-time lecturer), but it is not important to understand the detail of the example. The decomposition is dealing specifically with generating information PIs. The purpose of this example is to show that functional decomposition can be very specific to a computer-based system, as was the entity map for the college that we showed earlier. This functional decomposition is indented to provide the basis of computer programmes that will run the functions.

6.5.1 Double checking on entities and functions

Even the simplest of information problems will by now have generated quite a complex picture of the work to be done. When complexity increases it is useful to supplement our analysis with a little double checking to make sure that the picture of entities and functions we are developing is sensible. The best way to carry out such a check is to make use of an

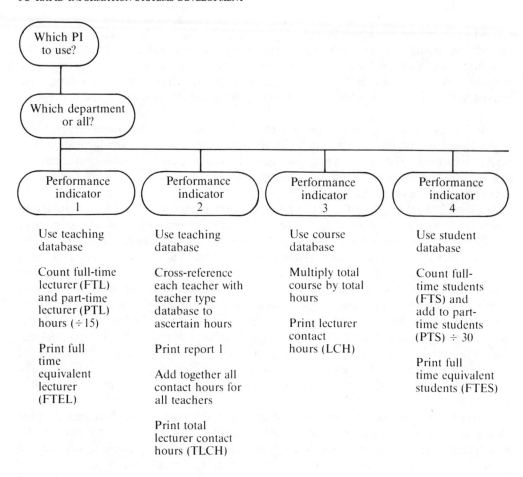

Figure 6.14 Performance indicator (PI) generation.

entity/function matrix. We have already mentioned entity/attribute matrices. An entity/function matrix operates on the same principle (see Fig. 6.15).

As with our previous example, the easiest way to demonstrate this is to give an example, and this is shown in Fig. 6.16 on page 100.

The rule is, if there are any functions without entities or entities without functions, or —more difficult to check—missing entities or functions, this tool should be able to pick up any problems.

This stage of the analysis is timed out as follows:

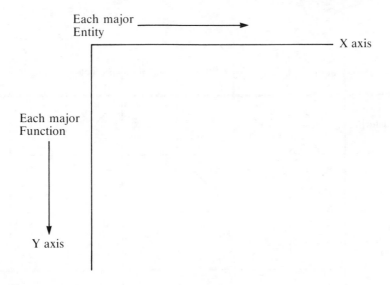

Figure 6.15 Entity—function matrix.

> **Amount of time devoted to analysis so far:**
> Total for this stage (functional decomposition) = 2 days
>
> Cumulative total to date = 12 days

It is estimated that this stage should take no more than two days.

Checkpoint At this stage in the analysis we have:

1. A clear entity model.
2. A correlation of the relationships of function to entities and also some understanding of how these functions decompose in the organization.
3. A listing of the key attributes of entities.

We can now go on to the next stage of the information modelling, which is to build in the events that will trigger the functions for the entities which have the attributes ('which lived in the house that Jack built!').

6.6 EVENTS

Here we will introduce another analysis tool that is very much from the 'hard' school of

Entities

Functions	Manage-ment	Accounts	Planning	Roads register	Regional and project data		
Administering	X	X			X		
Planning and Accounting	X	X	X	X	X		
Roads information collection				X	X		
Report assessment	X		X				
Evaluation of progress	X		X				
Receive programme proposal	X						

Figure 6.16 Entity—function matrix.

thought: the data flow diagram. This is, however, a useful tool for modelling the input of events that will trigger functions in the system.

Again you should realize that the issue of data flow diagrams is quite a major subject in the science of analysis and design and if you want to develop this aspect of the analysis for your own benefit we again refer you to the appropriate structured systems analysis and design texts in the further reading section at the end of this chapter. Considerable practice will be needed for a really detailed and profesional understanding. For now we will continue with our theme—an introduction to useful tools that can be applied rapidly.

The data flow diagram examines and demonstrates how information flows in functional hierarchies. Within the diagram the information flows from left to right through the functions. The functions are shown in boxes; events come in from the top (see Fig. 6.17). A key item to be aware of is the avoidance of ambiguity in terms of terminology, e.g. what 'entities'

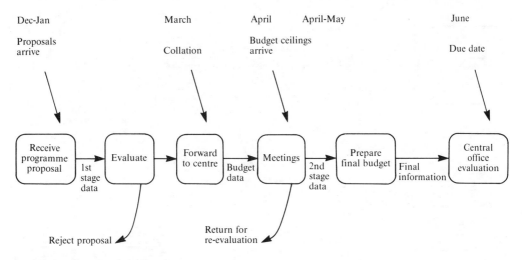

Figure 6.17 Data flow diagram.

behind the functions are referring to at any one time? In this case we are looking at a shortened outline of the events that trigger the functions in the planning and accounts major function as related to the management, administrtion, and planning entities. Ideally the data flow diagram will provide the analyst with another consistency check on the following:

- Where have events which trigger functions arisen?
- Do these events check with the realities of the present situation?
- Have you missed out any major functions in your analysis to date?

Generally, it is necessary to produce data flow diagrams of major or complex areas. It is not usually essential to model the events for all functions. In this case you see that it is useful (and possible!) to put concrete dates to the events.

In terms of the second phase of our rapid approach, we have completed the analysis when we have mapped out the events for the major functions that will form the core of our new information system, as mapped out in the conceptual model in stage 1. Stage 2 of our analysis can be set out in total as shown in Fig. 6.18.

At this stage much of the concrete detail of the eventual information systems has been produced.

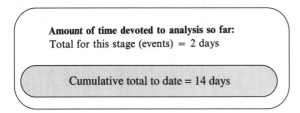

Amount of time devoted to analysis so far:
Total for this stage (events) = 2 days

Cumulative total to date = 14 days

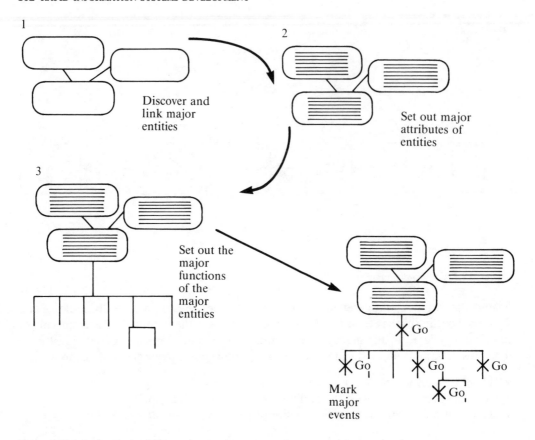

Figure 6.18 Information modelling.

A further two days should be sufficient to prepare the data flow diagrams *as and when necessary* (it is not envisaged that all aspects of the analysis will need this stage).

6.7 TYING IT ALL TOGETHER

The finalization of the information model should not take place without the overall schema being presented to all major stakeholders for agreement (Fig. 6.19). For this procedure the analyst will need to produce the information model in a form that is readily understandable to non-specialists.

It may be that the outcome of such a consultation process will be the need to rethink the conceptual model. The information model will provide a good basis for understanding the likely extent of the system. This in turn will have repercussions for the cost of the system.

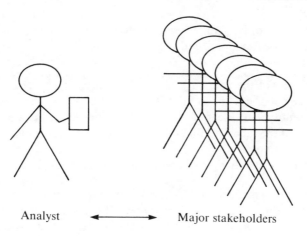

Analyst ←――――→ Major stakeholders

Figure 6.19 Information modelling requires agreement.

It may be necessary to think in terms of several phases eventually to produce the system that the model shows.

When the stakeholders of the system are content with the model, the third stage of the analysis can take place.

6.8 CONCLUSIONS

The information model provides hard details for the soft conceptual model. During this stage we may have to rethink our conceptual model. It may be that implications of the model are unworkable, too expensive, or too widescale in terms of their implications. This is all right. One of the purposes of this approach is that the information system can be adjusted in reaction to further study. Between human activity systems and information modelling there should be a feedback loop allowing for further refinement of the proposed system.

When the information model is complete, it can be put to one side. We need to keep reference of the amount of data and the number of entities that will need to be planned for. This information can be carried forward to the next stage of the analysis.

6.9 TUTORIAL

Exercise 3 On information modelling

At this stage we are assuming that you have compared the results of the exercises at the end of Chapter 5 with the model answer set out in Appendix 3. Having arrived at a conceptual outline of the new system, the crunch has come. We now need to turn our conceptual model into the outline information system. The way we shall do this is to set out major entities, attributes of key entities, functions, and triggers or events.

Your exercise is to:

1. Make an entity map of the company as described in your top-level conceptual model. Only focus on key entities. Try to keep it simple.
2. Related to your entity work, set out the major functions related to a major entity. Think about how many of the functions might be coped with by an automated system and how many of them could not.
3. What are the triggers/events that will be most important for the system.

Some hints Entity model construction. If there has been any discussion about the situation for regional offices, this must be put behind us now. The terms of reference are explicit and therefore we need now to break down the conceptual model systems into entities about which we wish to store information. Major entities might include planning and administration. These two will in turn be served by others. The top end of the entity spectrum is the issue of strategy for the company. Once you have drawn the entity model, take a look at the model answer in Section 4 of Appendix 3.

In listing attributes for the entities, only concern yourself with the primary items.

When working on the functional decomposition, focus on the key issue of strategy. Work out the preparation of strategy function with particular emphasis to the hierarchy of subfunctions that help to provide the strategy formulation necessary for effective management. A key issue here is understanding what the competition is up to. Feel free to go down two or even three levels in the decomposition. When you have completed this stage, take a look at the example worked out in Section 6 of Appendix 3.

When you come to look at a data flow diagrams, again focus on an element of the strategy formulation (e.g. the events noted which would contribute to the effective monitoring of major competitors).

When you have finished, take a look at Section 7 of Appendix 3.

FURTHER READING

Antill, L. and Wood-Harper, A. T. (1985) *Systems Analysis*, Made Simple Computer Books, Heinemann, London.

Ashworth, C. and Godland, M. (1990) *SSADM: A Practical Approach*, McGraw-Hill, London.

Avison, D. E. and Wood-Harper, A. T. (1990). *Multiview: An Exploration in Systems Development*, Blackwell Scientific, Oxford.

Avison, D. E. and Wood-Harper, A. T. (1991) Information systems development research: an exploration of ideas in practice. *Computer Journal*, **34**, no.2.

Bowers, D. (1988) *From Data to Database*, Van Nostrand Reinhold, London.

Lucas, H. C. (1985) *The Analysis, Design and Implementation of Information Systems*, McGraw-Hill, New York.

Nolan, R. L. (1984) Managing the advanced stages of computer technology: key research issues. In W. McFarlan (ed.), *The Information Systems Research Challenge*, Harvard University Press, Boston.

Wood-Harper, A. T. (1989) Comparison of information systems definition methodologies: an action research, Multiview perspective. Ph.D. thesis, School of Information Systems, University of East Anglia, Norwich.

7

TECHNICAL NEEDS, SOCIAL NEEDS—GETTING THE RIGHT BALANCE

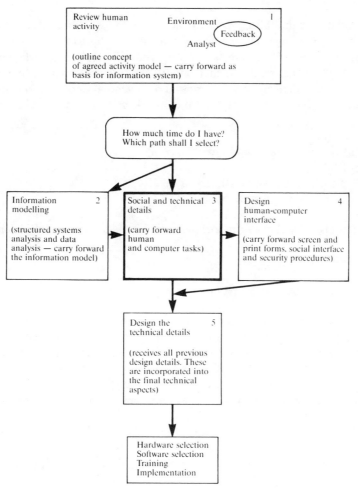

Figure 7.1 Rapid planning methodology.

Keywords socio-technical systems, future analysis, social objectives, technical objectives, social alternatives, technical alternatives, best-fit solution.

Summary This section covers the development of the implementation considerations of hardware, software, and people used to operate the outline system as depicted in phase 2 of the methodology. The integration of a variety of alternatives in terms of their costs, resource implications, and constraints leads to the expression of the actual tasks that will be accomplished by the various human and computer aspects of the system.

7.1 INTRODUCTION TO SOCIO-TECHNICAL SYSTEMS

It is not essential that the third aspect of the methodology should directly build off the information-modelling stage. The main reason for this relates to the various ways in which our approach can be used. If you are using path 3 you will not have undertaken information modelling at all. Nevertheless, if you have undertaken information modelling this will have provided us with an outline of the structure that our eventual information system can make use. We could even go so far as to say that the process will have identified for us a database structure:

Entity and entity types = files and records.
Attributes = fields.
Functions = menus of necessary actions to take.
Events = the triggers that prompt actions.

The items that might be of most importance for this stage of the analysis are:

- Details of entities.
- Details of number of attributes.
- Details of potential number of records.

If you do not have this information, do not worry. The third aspect of the total methodology can stand alone. It is a systems analysis and systems design process in itself.

Stage three, the design of the socio-technical system allows us to specify the nuts and bolts of the actual system itself in terms of human and computer tasks, human and computer requirements. After all, a system that is beautifully designed but completely inappropriate for the people who are available to use it or the environment that will support it is not much use at all. Therefore it is essential to consider the way in which people carry out their work, the vested interests and politics of the local situation, and the way in which the new system can best be fitted into it.

The process of the socio-technical design stage includes the outlining of:

1. Job design.

2. Specification of the human and computer (if appropriate) tasks.
3. Decisions about staffing and training requirements.
4. A detailed technical computer specification (if appropriate).

The analysis depends for its background context on the rich picture (especially for issues such as local power supply, availability of spares/servicing).

The job that the analyst is called upon to perform is to outline the various alternatives available to the stakeholders to provide decision makers with relevant and sensible plans for action.

The overall structure for the stage is shown in Fig. 7.2 overleaf. The outline set out in Fig. 7.2 demonstrates an eight-stage process:

1. Predict future environment analysis—this is the attempt of the analysis to build into any new information system some redundancy in terms of the system being able to deal not only with the issues of the present moment but the situation as it continues to develop.
2. Outline the social objectives and technical objectives—this stage sets out the general social needs of the system (improving job satisfaction, increased professionalism, etc.) and technical objectives (improving the timeliness of operations, holding and analysing data efficiently, etc.).
3. Outline the social and technical alternatives—the measures in the social and technical fields that can be taken to meet the social and technical objectives.
4. Generate by mixing the social and technical alternatives into different options. Here we show three. There should be at least two but there could be many more.
5. Rank the alternatives in term of their costs, resources, contraints, and occasionally benefits.
6. Select the best alternative mix—the 'best fit' solution. This is an important point. Our mix and match of social and technical alternatives to meet our needs will rarely appear to be ideal. We will select the best and not the perfect alternatives.
7. Work out the human and computer tasks to meet the best fit solution.

Taking this outline as our starting point, the analysis begins to develop as follows.

7.2 PREDICT FUTURE ENVIRONMENT ANALYSIS

To review social and technical resources and constraints for the development of efficient information systems without thinking about the future of such systems would be a short-term attitude. Of course, we cannot know exactly what the future will bring, but we can make some speculations on the nature of changes. The consideration of future conditions may well help us to avoid some of the problems identified in Chapter 1. The study of future environments and conditions is expressed in the 'future analysis' theory (Land, 1982, 1987). Essentially the analysis has four major foci:

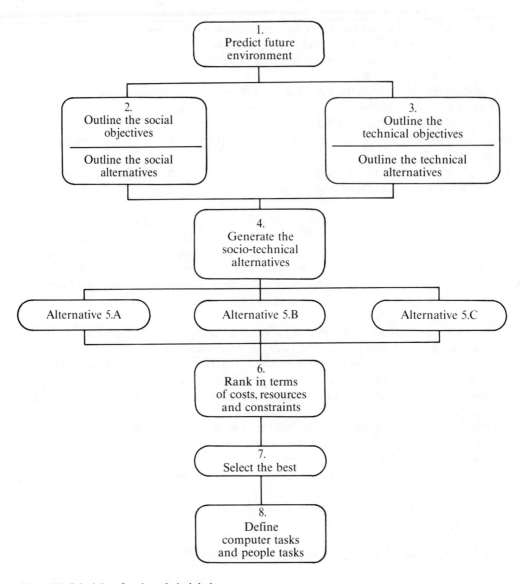

Figure 7.2 Principles of socio-technical design.

1. Prediction of the kinds of change that are possible—are they technological, legal, or economic? This requires us to look at the context and situation of the organization in which we are working and, possibly with the help of structure plans (if they exist), predict the mid-term development of the institution (3–5 year plan). This analysis stage should give us some idea of the type of expansion, contraction, and change that will occur and

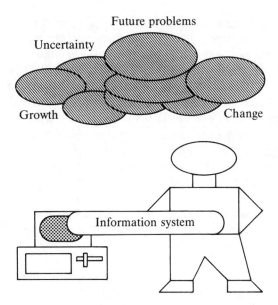

Figure 7.3 Predict future environment analysis.

which the incoming system will have to deal with.

2. The likely outcome of the system in the future—what will be the effects of an improved, possibly automated, information system? There are all kinds of disruptive and/or constructive events that may be related to the development of a new system (laying-off staff, employment of computer professionals, development of new grades to cover computer staff, etc.).

3. What features of the proposed system are more susceptible to change? Where would you expect the new system to change first? Can this be planned for? For example, will data collection procedures change? Will the existing departmental structures be maintained?

4. What is the extent and horizon of the system? This requires a long-term view (5–10 years). Obviously this is guesswork but it gives one a sense of humility in the initial design and requires us to speculate as to how what we plan today may be the building block for further developments into the long-term future.

The end of the stage is to set out the new system within the context of continuing development in the organization. The study allows us to see the system in the context of one moment in time.

Obviously this type of study can be quite detailed. Generally, however, it is quite a short procedure. In our case study of the governmental department (in brief) our findings were:

1. Prediction of the kinds of change that are possible. Major issue—fast growth for the department of the next five years; major investments in infrastructure (running to over

10 per cent of the department's annual budget). It is expected that the investment will lead to increases in employment within the department in terms of technically competent personnel. At a more political level, it is expected that the department will have a widening national responsiblity in terms of maintenance of existing roads.

2. The likely outcome of the system in the future. The greatest efficiencies should occur in terms of finance and staff administration. More specifically, there should be a reduction in unskilled staff employment and a tendency towards widening automation. The project is linked to related investment in improving communications.

3. What features of the proposed system are more susceptible to change? The assumption is that the core management will feel the changes first, which will cause fairly widespread disruption and prove unsettling for several years to come. This will require training and motivational support.

4. What is the extent and horizon of the system? Eventually the system will extend beyond the narrow centralized system to all the regional and district offices (again, this links to the communications investment).

Obviously the anlysis can be far more detailed if time and circumstance (and need!) demands. All subsequent analysis and design will need to keep in mind these considerations. They should be built into our planning and influence our choices in terms of the needs of the organization to cope with a new information system and the capacities of technology to cope with developments.

The next job is to agree and set the outline social and technical objectives of the system being created.

7.3 OUTLINE SOCIAL AND TECHNICAL OBJECTIVES

Social objectives refers to the expectations of employees and major stakeholders. The technical objectives refer to capacity of the organization as a whole to react to key issues. For our case we can set out the twin objectives as follows:

Social objectives

- To be relatively self-sufficient.
- To provide a quick service.
- To provide job satisfaction.
- To provide professional satisfaction.
- To improve the professional status of the department.

Obviously all these objectives are arrived at following consultation with all major stake-holders. This is a vital point. The social objectives of a system—broadly to be seen as the expectations of the system in terms of the human beings who are going to be working with it—will vary from site to site. No two information systems being planned for organizations will have the same objectives. Often the social objectives of a system are undervalued. Management do not tend to feel that the social needs of a system are as critical for system development as technical issues. Thinking back to the systemic/reductionist arguments of the earlier chapters, we cannot overstress the need for social issues to be adequately planned for. Having outlined the social objectives, the next task is to return to our conceptual model and information model to set out the technical objectives for the incoming system. In the case of our example these are as follows:

Technical objectives

- To inform management.
- To improve timeliness.
- To improve communication.
- To increase information-processing capacity.
- To provide a long-term facility.

These are the primary tasks we are hoping that the system will need to undertake. In our example the objectives are quite brief and broad. They could be very specific—as in the following example of one single technical objective of a management system for a large trading company:

> To provide daily commodity movement statistics—in the form of 34 agreed performance indicators on 15 key commodities for all managers above grade 10.

As we have said before, your own position will indicate the depth of detail you will need to go into. There may be a need for the analyst to grade objectives, particularly if interviews generate quite a lot of them.

7.4 GENERATE SOCIAL AND TECHNICAL ALTERNATIVES

The social alternatives refer to the description of different ways of organizing individuals to undertake the work required for the system, while at the same time achieving the social objectives. Technical alternatives should offer a range of means of meeting the information-processing requirements. The issue of technical alternatives does require the analyst to understand the basics of what each type of alternative might mean. For example, one view of the range of technical means to work an information system might be as follows:

1. An entirely manual system.
2. A manual system with computerizable aspects (e.g. those produced by the Kalamazoo Company).

3. A mixed manual system with some microcomputer aspects. Microcomputers installed on a **stand-alone** basis.
4. Information system based on stand-alone microcomputers.
5. Microcomputers **networked** together in a local area network (LAN).
6. Microcomputers networked onto a **minicomputer**.
7. A minicomputer system.
8. A **mainframe** computer system.

If you are not familiar with hardware and software options, do not worry—you are in good company. With the incredible continuing development of computer power it is almost impossible even for professionals in the industry to keep up with developments. As a very general briefing Table 7.1 provides an overview. Each option offers benefits and costs. You will need to consider in some depth what you are ready and able to recommend and analyse, and what the organization can cope with. In our example the alternatives worked out as follows:

Social alternatives

S1. In-house training of existing staff to provide the complete computer service.
S2. In-house staff trained to supervise new, pretrained staff.
S3. Trained new staff brought in.
S4. Use agency staff.

Table 7.1 General observation of systems

System	Selected features
Manual system	Limited processing
	Subject to basic analysis errors
	Well understood
Manual computerizable	New system
	Basic error avoidance
	Partially understood
Mixed manual/microcomputers	Old practice and new
	Basic software
	Thousands of records processable
	Reduction in analysis errors
	Increase in input errors
	Maintenance required

Table 7.1 Continued

System	Selected features
Microcomputers	New practice Basic software Thousands of records processable Reduction in analysis errors Increase in input errors Potential adoption problems Training needs high Maintenance required
Network	New practice Basic software Specialized software Thousands of records processable Reduction in analysis errors Increase in input errors Potential adoption problems Training needs very high Much maintenance required
Micro/minicomputers	New practice Basic software Specialized software Tens of thousands of records processable Reduction in analysis errors Increase in input errors Large potential adoption problems Training needs very high Sometimes require specialized environments Considerable maintenance needed
Minicomputers	New practice Basic software Specialized software Hundreds of thousands of records processable Reduction in analysis errors Increase in input errors Large potential adoption problems Training needs very high Often require specialized environments
Mainframe	New practice Basic software Specialized software Millions of records processable Reduction in analysis errors Increase in input errors Large potential adoption problems Training needs very high Only available in specialized environments Considerable maintenance needed

The social alternatives are ranked from S1 to S4. Essentially we are looking at ways of operating a computer-based system. The options range from the use of existing staff, trained in the skills required, to the use of agency staff only. The technical alternatives are as follows:

Technical alternatives

T1. Minicomputer and terminals.
T2. Microcomputer network.
T3. Microcomputer, stand-alone.

Note that we have not included in our logical alternatives the possibility of using a manual system. This is because the alternative was not relevant following discussions with the funding agency. This is a real-time limitation on any analysis and again reiterates the importance of the terms of reference on analysis and design routines.

7.5 RANK THE ALTERNATIVES

The next stage of the analysis is to mix and match the various alternatives together and then to list all the alternatives and reject any that are immediately not feasible. For example:

S1 T1—reject outright on grounds of cost
S1 T2
S1 T3
S2 T1—reject outright on grounds of cost
S2 T2
S2 T3
S3 T1
S3 T2
S3 T3
S4 T1
S4 T2
S4 T3

Of course you can be much more sophisticated than the example we have here. For example, you may have set out the intial alternatives so that that multiple social and technical combinations could be selected:

S1 = existing staff
S2 = trained existing staff
S3 = new trained clerical staff
S4 = new trained management staff

T1 = microcomputers
T2 = network
T3 = micro/minicomputers

And then the mix might be:

S1, S3, T1
S2, S4, T1, T2
S3, S4, T3

and so on.

Assuming that we are working from our first example, we now need to find the optimum combination of alternatives. We need to consider each pairing in terms of their implications for *cost, resources, constraints* and if needed (though not usual), *benefits* (if thought necessary). These factors can be assessed using one of three methods.

1. They can be arrived at by a process of cost/benefit analysis. This can be a lengthy process.
2. They can be set out on a table and graded against each other on a ranking of 1–9. For example, we set the number values as follows:

1 = very good
2 = good
3 = quite good
4 = better than average
5 = average
6 = worse than average
7 = quite poor
8 = poor
9 = very poor

Our alternatives could then be set against each other as follows:

	Social costs	Technical costs	Social resources	Technical resources	Social constraints	Technical constraints	Total
S1T1	4	4	3	3	1	1	16
S3T1, S4T1	8	8	7	8	6	9	46
etc.							

The idea being that the alternatives with the lowest totals would indicate the more appropriate choices. This procedure can also be quite time consuming.

3. An even simpler rule of thumb would be just to set out the major factors against each aspect. For example:

S1T2 In-house staff and microcomputer network
Social costs—training in use of software and network.
Technical costs—software, hardware.
Social resources—some trained staff.
Technical resources—some hardware.
Social constraints—none of real importance.
Technical constraints—power supply, climate.

S3T1, S4T1
Social costs—high cost of new staff, training in use of software and network.
Technical costs—very expensive software, hardware.
Social resources—none.
Technical resources—none.
Social constraints—introducing new staff to practices.
Technical constraints—power supply, climate.

By using this method we can rapidly peel off the most expensive alternatives. We might still have trouble however, if we have very close alternatives to compare. In this case we might feel justified in falling back on the second method of assessment.

The result of this phase will be a ranking of the alternatives in order. Our ranking from our exercise is as follows:

1. S1T2.
2. S1T3.
3. S2T2, S2T3.
4. S3T2, S3T3, S4T2, S4T3.
5. S3T1, S4T1.

Our best mix is therefore, the use of trained internal staff and a microcomputer network.
   For the purposes of this exercise we will work with this combination and, in applying the limitations that it will inevitably present, attempt to relate it to our information model arrived at in the last section.

7.6 HUMAN AND COMPUTER TASKS

The final stage of the socio-technical specification is to outline the *people tasks* and *computer tasks* for the new system.

7.6.1 People tasks

The people tasks that you set out must be capable of dealing with the wide range of issues and potential problems you have thought of in your future analysis, as well as the range of data tasks that were implied by the information-modelling carried out in phase 2 of our approach. These can be broken down along a number of lines but for our purposes we provide a four-way division into management tasks, input–output tasks, training tasks, and maintenance and support tasks.

In general terms we might structure these as follows:

- Management tasks:
  - Overall management of the system.
  - Management of an effective reporting procedure.
- Input–output tasks:
  - Data input to the system.
  - Selective output from the system.
  - Report generation.
  - Interpreting the output.
- Training tasks:
  - Training senior management in familiarization.
  - Training assistant managers and administrators in use.
  - Training clerks in operations.
  - Training technicians in information technology maintenance.
- Maintenance and support tasks:
  - Repairing faulty items.
  - Regular servicing.

Several points can be made concerning this type of checklist. Firstly, training should always begin with the top management. This ensures that the system will be supported. One problem that is constantly recurring is the potential alienation of managers through assuming their commitment without training. Information technology is generally seen as being threatening to those who are on its margins. If a system is planned and senior staff are not given strong support in its uses and values, it can happen that the system will be underused by other staff and will lack the political support to gain thorough acceptance. Secondly, the outline that we prepare now will not be worked out in detail at this point. It will be the task of a final stage of our approach actually to set out the major aspects of the configuration in its final form. Here we are attempting to provide ourselves with the overall guidelines for the coming system. Thirdly, we need to gain the assent of all major stakeholders at this stage to the tasks we outline. All such tasks will have immediate and recurrent costs for the organization and the donor. The analyst may need to reduce his or her expectations in the light of what is financially feasible. This point also is true for the next stage, computer tasks.

7.6.2 Computer tasks

These will tend to be rather easier to structure at this level. Generally two levels need to be considered: the data and the equipment.

The data This refers to the actual items that need to be accomodated. The type of data that our government department needs to be stored and retrieved is as follows:

- Roads register data.
- Heavy equipment inventory.
- Employee records.
- Project data.
- Accounts data.
- Other items (miscellaneous), e.g.
  - word-processed documents,
  - spreadsheet matrices,
  - database files,
  - graphics files,
  - road design files.

The equipment At this stage the equipment prescribed can be set out by general function rather than actual hardware and software. Thus, with our department we could specify:

- Networking of data to key personnel (approximately 16 units)—multiple file access to root files.
- Quality output as required.
- Draft output as required.
- Equipment to deal with power fluctuations and power down (potentially hours).
- Capacity and facility to archive (approximately 40 megabytes in year 1).

Amount of time devoted to analysis so far:
Total for this stage (socio-technical) = 6 days

Cumulative total for paths 1 and 2 = 20 days
Cumulative total for path 3 = 13 days

This stage could be completed in six days.

7.7 A SECOND WORKING EXAMPLE—A UNIVERSITY INFORMATION SYSTEM

For your consideration we include here a total socio-technical analysis for a much simplified university management information system.

7.7.1 Predict future environment analysis

This is very subject to error. The task requires a look into the crystal ball on four levels:

Predict changes possible? These can be listed as follows:

- National authority required to handle funding for regional as well as national universities.
- Growing number of universities.
- Increasing need for good planning information, e.g. sophisticated budgetary information (fiscal planning).
- Increasing student population.

All the above factors produce a growing pressure on any incoming system in terms of disk space, processor speed, rapid response.

- Political uncertainty (general election imminent).
- Danger of loss of skilled staff to the private sector.
- **UNIX** minicomputer system.

The above indicate the extreme flexibility and volatility of this situation. The section can be closed by stating that the net result is to indicate a situation of rapid change and high uncertainty. In the light of this it should be recognized that any incoming system is initially (at least) vulnerable.

What is the likely outcome of the proposed system in the workplace?

- Increased information for planning.
- Increased timeliness of information.
- Increased access to key information.

Therefore, the opportunity exists for increased efficiency or for institutional paranoia. This worry created by any new system is most likely to be felt in the financial area. In particular there is evidence that conflict may well arise between university financial managers and academic planners vying for the same information for decision autonomy.

What aspects of the new system are most likely to change first?

- Positive:
  – The information demanded will develop and grow in complexity.
  – Managers will become increasingly reliant upon rapid information to support decision making.
- Negative:
  – Information systems will increasingly be seen as having to be computer based.
  – Computer-based statistics trusted without due care of data validation or verification.
  – Redundancy of the hardware and software.
  – Uneven national development between universities.

What is the long-term extent of the system?

- Wide area network (**WAN**).

7.7.2 Social and technical objectives of the system

Social objectives

- Improved information service.
- Improved professionalism.
- Improved funding choices.

Technical objectives

- More accurate information, more rapidly.
- Internal self-sufficiency in maintenance.
- Initial system viable for future growth.

7.7.3 Social and technical alternatives. Ways to achieve the objectives

Social alternatives

- Train internal staff—S1.
- Train internal staff and employ new—S2.
- Employ new staff—S3.
- Use contract workers plus internal—S4.

Technical alternatives

- Stand-alone **DOS** systems—T1.

- Computer network—T2.
- UNIX system with DOS applications—T3.
- UNIX system—T4.

7.7.4 Ranking the alternatives

As a rule of thumb in situations where there is very little time to carry out a detailed analysis (e.g. cost/benefit analysis of each alternative), we assume a scale of 1–9, where 1 is very good and 9 is very poor. The ranking covers the analyst's perceived understanding of how each alternative scores in terms of its costs, resources available, and constraints within the system. Usually all the details for each alternative are argued out. In Table 7.2 we deal only with the details of the top three alternatives. It should be remembered that the anlysis is taking place against the background of our future analysis in Section 7.7.1. This assumes certain long-term trends which make simple, single-user systems on DOS very redundant in the longer term. The analysis should ideally be carried out for the short, medium and long term. Here we are assuming certain long-term trends identified in Section 7.7.1 but are focusing on the medium term.

Table 7.2

	Social costs	Technical costs	Social resources	Technical resources	Social constraints	Technical constraints	Total
S1T1	2	7	3	3	3	9	27
S1T2	6	8	7	8	6	7	42
S1T3	6	8	6	8	6	6	40
S1T4	9	8	8	8	8	8	49
S2T1	4	7	4	5	3	8	31
S2T2	5	6	5	6	3	5	32
S2T3	5	6	5	6	3	4	31
S2T4	6	9	8	7	5	3	38
S3T1	7	7	7	2	7	6	36
S3T2	7	6	7	4	7	4	35
S3T3	7	7	7	7	7	3	38
S3T4	7	9	8	7	7	3	41
S4T1	9	6	9	2	8	6	40
S4T2	9	6	8	4	8	4	39
S4T3	9	7	8	5	8	2	39
S4T4	9	9	9	7	8	3	45

Social costs = introducing new staff, learning new skills, fitting in, etc.
Technical costs = learning new systems, supporting new systems, redundancy of old systems, replacement costs, etc.
Social resources = high levels of awareness, low number of skilled staff, high reluctance, etc.
Technical resources = existing equipment, existing software skills, no existing equipment, etc.
Social constraints = acceptable work practices, managerial support, etc.
Technical constraints = long-term viability, organizational capabilities.

The analysis, however imperfect and subjective, indicates three options which fit one total implementation pattern given the future analysis. The pattern is as follows:

- S1T1 which leads to:
- S2T1 which leads to:
- S2T3.

This might be seen as constituting a development plan:

S1T1. Initially the system comprises internal staff using stand-alone computers. This is low risk in terms of investment and discouraging staff. The situation changes when staff have been made familiar with MIS concepts on the new equipment by bringing in new, trained staff ready to adapt to the third phase.

S2T1. The second phase can be seen as a period in which existing staff and new staff get to work together.

S2T3. Phase three brings in the UNIX network but using, in the early stages at least, existing DOS applications. The new staff should have some UNIX awareness and this will hopefully be transferred to existing staff. In the long term there is the capacity to upgrade to a full UNIX system.

7.7.5 Define computer systems and human systems

Human tasks The incoming management information system requires the development of a range of skills, including:

1. System needs.
   (a) Network management.
   (b) System management:
      - System security
      - Hardware maintenance.
   (c) Computer hardware architecture.
   (d) Database design.
2. Management needs.
   (a) Project management.
   (b) Analysis and design.
   (c) Project implementation.
   (d) Monitoring and evaluation.
   (e) Principles of data collection, validation, and verification.
3. Operational needs.
   (a) Supervisory.
   (b) Keyboard entry.

Some of these items are already in evidence. Training will need to reflect these areas. In particular each computer unit will require:

1. A computer unit manager to deal with day-to-day running of the system.
2. A computer unit technician to deal with preventative and corrective maintenance.
3. Several clerks for data input. There will be a seasonal fluctuation in this activity and at times the consultants believe there will be a considerable flow of data for the system.
4. Ideally each unit should also possess a systems analyst and a programmer to make alterations and pursue software problems as they arise.

Computer tasks Any incoming system will have to deal with some considerable number of records (e.g. there are approximately x thousand students in the universities being taught by approximately x hundred lecturers). The database forms will be quite large and considerable storage will be needed in order for data analysis to take place comfortably. Any incoming computer system will need to be able to cope with databases of 50–60 fields, possibly related to several other databases at the same time, potentially containing thousands of records.

7.8 CONCLUSIONS

The example set out in Section 7.7 is taken from a practitioner's notes. There are various points that it demonstrates and which we would like to express in general conclusions.

1. At this stage you do not need a detailed knowledge of the technology. You can recommend in long hand the types of things you need to do. This can be sorted out later on either by yourself and numerous trade magazines or with the assistance of computer sales companies (more on how to deal with these characters later on).
2. You can arrange the alternatives in any type of order. Analysis will be subjective.
3. Always bring the stakeholders into this stage of the analysis. You will need to explain the socio-technical combination that you finally recommend.

Following this stage, depending upon the path you are using you are either ready to go on to the human-computer interface or technical subsystems.

7.9 TUTORIAL

Exercise 4 On socio-technical systems

We have outlined the development of the social and technical systems that will make up our final system. Your task is to do the same.

1. Working from the example you have to date, set out the range of social and technical alternatives that will meet the information system you have modelled. Remember your terms of reference and your budget.

2. Work your way through the analysis. Give reasons for arriving at the conclusions you have.

Some hints One way of simplifying the social and technical objectives outlined is to set out some of the larger social objectives (as first set out in the CATWOE stage) and set against each the related technical objectives, e.g.:

Social objectives	Technical objectives
Improve planning	Technical skills
	Automated features
	Networking decision making

The alternatives arising from the objectives might be fairly standard (e.g. ranging from retraining staff to adding new staff on the social side and from a manual system to networked microcomputers on the technical).

The resulting ranking is probably best carried out by using the table method—comparing costs, constraints, and resources. It would be best to set out all the details of the table here.

When you have finished compare your answers to those in Sec. 8 of Appendix 3.

FURTHER READING

Avison, D. and Wood-Harper, A. T. (1991) *Multiview: An Exploration in systems development*, Blackwell Scientific Publications, Oxford.

Booth, P. (1989) *An Introduction to Human-Computer Interaction*, Erlbaum, Hillsdale, NJ.

Land, F. (1982a) Adapting to changing user requirements. *Information and Management*, **5**, pp. 59–75.

Land, F. (1982b) Notes on participation. *Computer Journal*, **25**, no.2.

Land, F. (1987) Is an information theory enough? In Avison *et al.* (eds), *Information Systems in the 1990s: Book 1—Concepts and Methodologies*, AFM Exploratory Series no. 16, Armidale NSW, University of New England, pp. 67–76.

Land, F. (1991) Information systems domain. In R. D. Galliers (ed.), *Information Systems Research: Issues, Methods and Practical Guidelines*, Blackwell Scientific, Oxford.

Mumford, E. (1981) Participative system design: structure and method. *Systems, Objectives, Solutions*, **1**, pp. 5–19.

Mumford, E., Hirschheim, R., Fitzgerald, G. and Wood-Harper, A. T. (1985) *Research Methods in Information Systems*, North Holland, Amsterdam.

8

THE HUMAN—COMPUTER INTERFACE

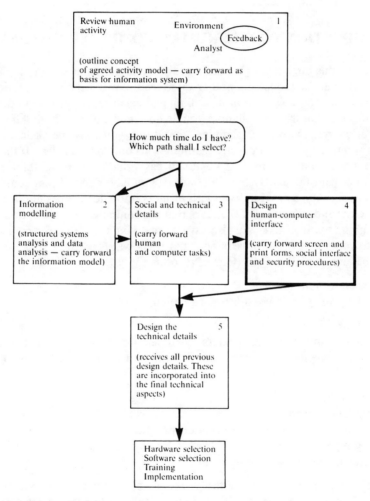

Figure 8.1 Rapid planning methodology.

Keywords technical interface, social interface, security interface, design procedure, menu-driven systems, priority access.

Summary It is essential that the user be able to work and communicate effectively with the computer. This chapter works out some of the details of the tasks discussed in the previous chapter. Various potential problem areas are reviewed, including work styles, sources of discontent in terms of new work practices, **dialogue systems** (the way in which computers communicate with users), and the security of different user groups. Security can be seen as a way of protecting the system from the user as well as a means of safeguarding valuable and sensitive information.

8.1 INTRODUCTION TO THE HUMAN–COMPUTER INTERFACE

If you are using this book to assist you in analysis and design as set out in Chapter 4 you will have arrived at the human–computer interface following the three previous stages of human activity system, information modelling and socio-technical system design. By this stage you may feel that there is still more you do not know rather than do know concerning the information system you are planning. One thing you must be clear about by now, however, is whether the system you are planning will contain computers. If the system does not require a computer system, this stage still has value in terms of thinking about the impression that your manual system will have. If you have decided that computers will be required, then this stage provides you with insights into effective design. We assume that you have a clear understanding of the tasks that the computer and the various users in the system are going to undertake. We also assume that you know that the computer systems are going to communicate with users and that users will need to be assisted in what can appear to be an unequal struggle. Our next job is to:

- Explain what an interface is.
- Look at the principles of good interface design.
- Review examples of design.

The human–computer interface refers to the environment in which the user and the hardware come together to perform the information system operations. The range of functions that can be carried out include:

1. The input of data.
2. Checking data for errors.
3. Making enquiries concerning information products.
4. Producing reports at certain *events* (remember our information modelling?).
5. Management, security, and monitoring.

However, the manner in which these tasks take place and the soical implications involved in different work practices need to be understood if the system is to work well. Key issues

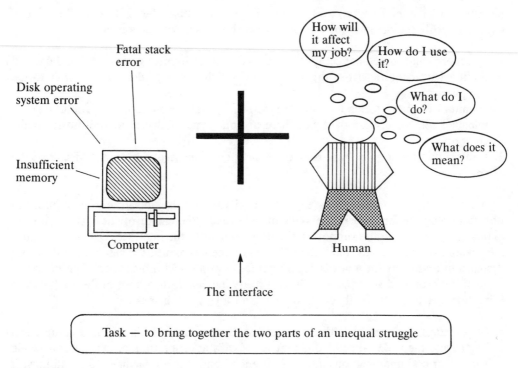

Figure 8.2

pertaining to work practice should have already been described in the human activity system stage in Chapter 5. If they were significant, they will have also been structured in terms of alternatives in the socio-technical stage in Chapter 7. Quality of work life is an important aspect of the total design. The interface set up here needs to provide end users not only with a technically sound system but also with a work situation that maximizes work interest at the same time as minimizing the negative aspects of technology (de-skilling, removing job satisfaction).

Therefore, when dealing with the human–computer interface we are dealing with a wide range of issues, such as those set out in Fig. 8.2 above.

Keep in mind one golden rule: you are designing systems for users not for computer experts. If your system is to be used, it has to be not so much 'friendly' as recognizable and useful to those who are going to use it. This may seem a simple idea but it is one that appears to be profoundly elusive to 80 per cent of those involved in information systems planning.

8.2 THE NATURE OF THE INTERFACE

At root the computer has to be able to deal with any of the range of tasks that have been identified in Chapter 7 and give a response that is understandable and useful. For this to occur smoothly the designer must plan around three technical facts of life:

1. Keyboards and sometimes computer **mice** are used for inputting data, editing data, and dealing with enquiries from the user. These are the primary devices for input to the computer.
2. Monitors in many forms (colour, monochrome, high resolution, low resolution) are intended for the display of data. These are one type of device for output from the computer.
3. Printers and plotters for **hard copy** of reports, etc. The second output device from the computer.

Depending upon your situation (e.g. the capacity of your organization to fund information technology related work, the skills of colleagues, the extent of the system you are planning), you will require support in the development of the interface. Usually programmers would set up the nuts and bolts of the interface mechanisms. These mechanisms are designed in outline by the analysis in an interactive process with the range of ultimate end users of the proposed system. As we have already indicated, the object of the analyst is to bring together two apparently contradictory and divergent themes:

- To provide the computer with clear, concise, and unambiguous data and commands.
- To provide the wide range of different users with appropriate interfaces which, while being clear and unambiguous, are also 'friendly' and provide facilities for dealing with mistakes and accidental entries as well as the correct ones.

The latter theme involves the second, social, aspect of our system. In outline this means the way in which we deal with the potential problems arising from what is often seen as a confrontation between the users and the computer. One starting point is to look at existing communication forms (standard data collection forms, reports, etc.) within the organization and attempt to copy those that are most appropriate. The purpose of the exercise is to allow the user and the computer to 'understand each other'. To make this possible from the user's point of view we will need to try to faciliate understanding of what is going on inside the 'box'. Much initial user training is often required in this area.

8.3 ISSUES IN JUDGING THE INTERFACE FOR THE USER

8.3.1 The technical interface

Systems can only be flexible up to a point. It is not possible to have a system that is adaptive to any type of user that happens to come along. However, this is not usually required. The analyst is normally involved in looking at certain types of user, their basic requirements,

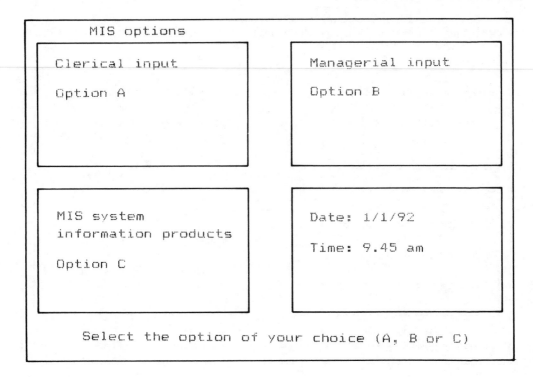

Figure 8.3

and then trying to match (by a process of interaction and participation) the right interface for the right user.

It is not difficult to think of examples. Clerks will have very specific, fixed needs from the incoming system. They will generally be familiar with existing manual input to registers, ledgers, etc. The basics of these operations will be well known and so our interfaces can be matched to what is already understood.

Researchers, managers, or planners, on the other hand, are likely to want to talk about targets, aggregates, trends, etc. They will want to be able to manipulate large volumes of data quickly and may well be uncomfortable using limited interfaces. Obviously the sorts of interfaces you plan for these different categories of staff ought to reflect those differences. An example will demonstrate some of the problems that we regularly encounter with a monitor display. Figure 8.3 shows a representation of the initial monitor input screen to a management information system. As you can see there are three key options—Clerical, Managerial or MIS products. Put yourself in the position of the users of this system. Obviously three broad bands have been planned for:

1. Clerical workers inputting data.

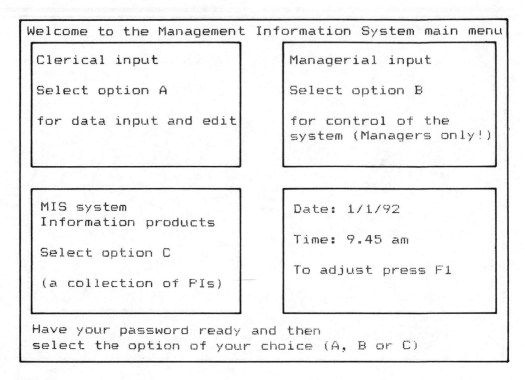

```
Welcome to the Management Information System main menu

  ┌───────────────────────────┐   ┌───────────────────────────┐
  │ Clerical input            │   │ Managerial input          │
  │                           │   │                           │
  │ Select option A           │   │ Select option B           │
  │                           │   │                           │
  │ for data input and edit   │   │ for control of the        │
  │                           │   │ system (Managers only!)   │
  │                           │   │                           │
  └───────────────────────────┘   └───────────────────────────┘

  ┌───────────────────────────┐   ┌───────────────────────────┐
  │ MIS system                │   │ Date: 1/1/92              │
  │ Information products       │   │                           │
  │                           │   │ Time: 9.45 am             │
  │ Select option C           │   │                           │
  │                           │   │ To adjust press F1        │
  │ (a collection of PIs)      │   │                           │
  └───────────────────────────┘   └───────────────────────────┘

 Have your password ready and then
 select the option of your choice (A, B or C)
```

Figure 8.4

2. Managers for controlling the system.
3. Information consumers needing to know things.

Complaints about this interface were as follows:

1. General—too vague. Users would like to know more about what option they want. 'Tell us what we want!'
2. Clerical comments—generally OK.
3. Managerial comments—resentment about sharing a system, would rather be physically unconnected to the clerical and information product aspects, need less guidance and menus and more control.
4. Information users—the system throughout gave information that was preplanned, there should be less 'patronizing' menus and more capacity for users to formulate questions.

Of course some of these comments go outside the scope of this single menu display, but it does go to show that even a simple introductory menu can have problems.

One attempt to improve the system is shown in Fig. 8.4. The main criticism of Fig. 8.4 was that the designer had gone too far. This screen is too full, tries to say too much, and

```
The Management Information System

Press

A for clerical input

B for managerial control

C for PIs
```

Figure 8.5

still does not segregate out managerial from clerical functions. Possibly a better approach would be to start off by asking each user if they want one of the three options, as shown in Fig. 8.5, and then proceeding with the relevant options for each group.

However, variations in what users require from the interface should be seen as only part of the dynamic operating between the two aspects (human and technical) of the proposed working system.

8.3.2 Social issues

Automation will tend to change radically the nature of work itself, often concentrating decision making in the hands of a few managers while simplifying work procedures for other groups.

Willcocks and Mason (1987) have attempted to structure the impact of technology on jobs by use of the matrix shown in Fig. 8.6 overleaf. The matrix demonstrates the manner in which a variety of jobs will lose skill content because of higher levels of automation (the assumption being in this case that cleverer computers and systems are taking the place of an existing installation). Anecdotally, from our own experience we might argue that the reverse is also true in certain cases (see Fig. 8.7 on page 133).

Therefore, individuals can make use of incoming technology to improve their situation within the organization. Whether information systems can lead to job expansion or contraction, the overall need is for the social impact of computer systems to be understood prior to installation. Willcocks and Mason argue that a series of implications follow automation:

1. The division of mental and manual tasks.
2. Maximization of specialization in skills.

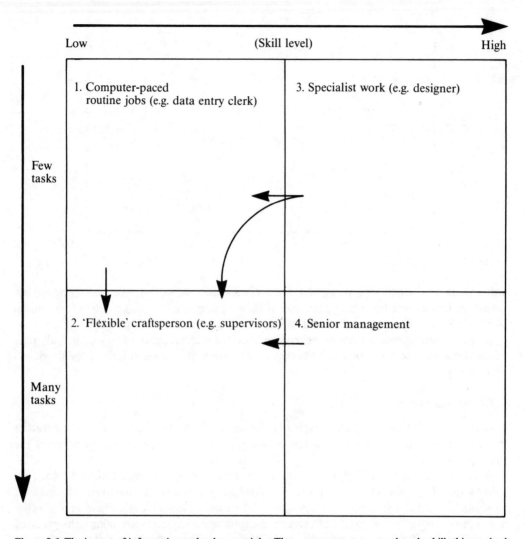

Figure 8.6 The impact of information technology on jobs. Theory: computers can tend to de-skill; this can lead to hostility and reluctance. (Adapted from Willcocks and Mason, 1987.)

3. Minimize skill requirement.
4. Minimize training and learning time.
5. Achieve full work-load.
6. Minimize variety of work.
7. Create short-cycle, repetitive jobs.

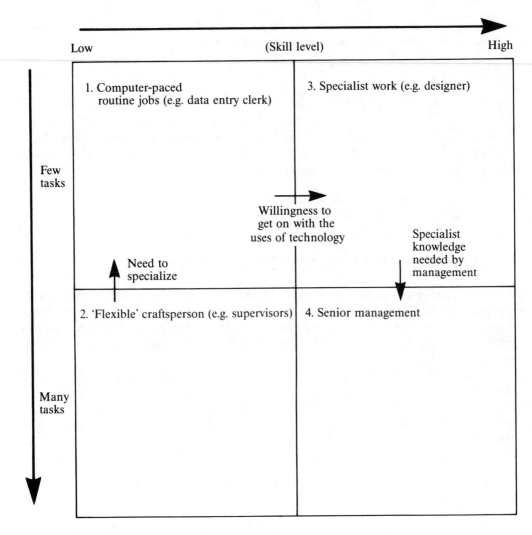

Figure 8.7 The alternative impact of information technology on jobs—adapt or die!

To determine whether all or part of this list of implications occurs requires the analyst to consider the human–computer interface in terms of job interest and job content.

There are a number of ways in which de-skilled work and relatively dull computer work can be improved. These include:

- Goal-orientated working procedure—whereby the staff are given a sequence of short-, mid-, and long-term goals to enhance job interest.
- Working teams—to share a series of low interest tasks.

- Vertical work groups—to distribute a range of working procedures around a group, rotating the higher value tasks to encourage interest.

8.3.3 Security

Another interface issue is that of security and the issue of user exclusion from certain aspects of the automated system. The issue of exclusion arises because *an inexperienced, unauthorized, or malicious user allowed free access to a computer-based information system can cause intentional, wilful, or unintentional harm to that system*. Again, two apparently contradictory positions have to be reconciled:

1. The need to allow users to make the most of the system.
2. The need to protect the system from the user.

8.4 DESIGNING THE TECHNICAL INTERFACE

At this stage we are designing the overall system and should not be too concerned with implementation and the problems that may well arise at that stage. It is quite important that we feel free to design the 'right' system rather than the one that is most practical. The two things may be identical or it may be that we will need to work gradually towards the system you believe to be best.

The provision of information to users requires us to look at the manner of the dialogue that will take place. Willcocks and Mason (1987) suggested a range of methods for implementing dialogue. These methods are shown in Table 8.1. The table shows at least 12 ways in which the technical interface between user and machine can be handled. In our example in Section 8.3.1 the second option—'Menu List'—was selected and various problems arose with this. It is quite possible to have several different interfaces for different

Table 8.1 Methods of implementing dialogue (adapted from Willcocks and Mason, 1987)

1.	Form filling	Form displayed; blanks to be filled by user
2.	Menu list	A set of options displayed, user keys in choice
3.	Question and answer	Computer displays a range of questions
4.	Question and restricted answer	Computer displays questions, restricted vocabulary available to answer
5.	Question with three possible answers	As above but only 'yes', 'no', and 'don't know' available
6.	Function keys	Input restricted to keys made available
7.	Natural language	Dialogue conducted in user's natural language
8.	Command language	Well-defined, limited number of commands available
9.	Query language	User request expressed in specified language
10.	Hybrid dialogues	Mix of dialogue styles available
11.	Parallel dialogues	Two or more dialogue styles available at user request
12.	Graphic interaction	Conveys complex interactions graphically

types of user. For example, in Sec. 8.3.1 we identified three different system users—clerks, managers, and information consumers. A single **menu-driven system** was planned. Possibly a better system might have been to provide:

1. Clerks with the highly controlled menu system.
2. Managers with natural language so that very precise commands can be given.
3. Information consumers with question-and-answer facilities.

Of course each of these options requires separate programming and therefore this type of techical interface design is quite costly.

The first point in any interface design exercise is to be sure that you have identified the range of users working in the system. In the case of the example shown in Sec. 8.3.1, method 2 was universally adopted despite the problems identified by the users. The reasoning behind this approach was that menu-driven systems are one of the safest methods. To get around potential problems with three user groups being supported by one system, the menu system was supplemented with method 11 because there were a variety of user groups, each needing a distinct information product from the dataset.

Another point to keep in mind when selecting the right interface for the right user is that any specific operation will involve the system in a number of exchanges. For example, to provide an administrator with staff information involves several different dialogue exchanges. For example:

- A review of the staff/division listings.
- Determining whether a required individual is still employed.
- Confirmation of department.
- Confirmation of priority to look at sensitive material.

And this structure does not take into account the problems of accident, incomplete information listing, forgetting the staff member's registration number, or incorrect spelling of his or her surname. Without becoming unnecessarily concerned, the analyst must realize that the potential for errors to arise in any system are almost countless. Systems should be provided with facilities to deal with such issues as:

- Incorrect initial data entry leading to problems with search and retrieval.
- Incorrect retrieval information.
- Insufficient data available for specific request.

At this stage it is worth noting where potential errors can enter the system (e.g. basic data entry by clerical workers), the type of expected error (e.g. misspelling, double entry), and the type of error checking that will be required (e.g. random checking of spelling—one record per one hundred, computer check of record to see if it already exists, etc.) and the action to be taken (e.g. prompt to edit).

8.4.1 The case example: specific issues

In the case of our government department there were a number of special items that had to be covered in our analysis. An initial concern was that almost all computer software is written in the English language. In our case, although much of the internal information in the department is in English, much of the input and the output is produced in local languages. Moreover, many of the junior staff have only a very limited grasp of English. However, following a further stage of analysis we were satisfied that the system would mainly be used by senior staff who tended to work most closely with the English language and that the manual system would run side by side with the incoming system, so the facility to support information integrity (in case of loss or damage) would be available from that manual system. The junior staff acting as operators on the incoming system would be selected in terms (partly) of their good grasp of English. So we could design our dialogue working on the valid assumptions that:

- English is the language medium.
- There will be various levels of interaction.
- Interfaces will need to be designed for each level.

The levels of user we identified were as follows:

1. Senior managers.
2. Computer unit staff.
3. Clerical staff.

Our next problem was to set out the principles of the user interface most appropriate to each user level. Computers tend not to be too friendly in this area. For example one very standard type of microcomputer system on sale throughout the world would give you the following message:

BDOS ERR ON B:

if there was a problem with your storage media. This is not a very informative message. Today things are improving but basic systems still come up with messages like:

FATAL STACK ERROR
ERROR 163

And even worse you may be asked to deal with the problem when you see the message:

Abort, Retry, Ignore?

not knowing what the result of any particular action in the context of the problem will be. For our three groups we now need to think about the types of dialogue which will be most appropriate.

The senior managers This group should not require a detailed knowledge of the system as

```
DEPARTMENT MIS


Choose your information source.

1. The roads register
2. The heavy equipment inventory
3. The register of projects
4. The accounting system
5. The administrative system
6. The urgent requests system


Q. To quit the system

Please type either 1, 2, 3, 4, 5, 6 or Q.
```

Figure 8.8

such. They will require information 'summaries' upon which they can either formulate further enquiries or make decisions. With this brief in mind we specify a design for the interface to be used:

1. Menu driven.
2. Access via code number.
3. Summary screens of data.
4. Urgent request facility.

We will go through the details of this design stage by stage, but it is useful to explain the basis of a menu-driven system in a little more depth. For most of the specialized work carried out on the databases of information it will be possible to pre-write menus of commands from which the user can select a function. For example, on logging in, a senior member might see something like Fig. 8.8. Here the manager is not required to know any more than which function he requires. This menu will lead to another and so on.

Item 2, the code number, refers to the security of the system and the issuing of code numbers that are to be confidentially held by all users. We will be looking at this area in depth in Sec. 8.6.

Computer unit staff This includes all individuals who are working directly with the computer system in terms of its day-to-day running and operation. The group might include:

1. The computer manager(s).
2. Computer advisers.
3. Computer trainers.
4. Computer operators.
5. Computer technicians.

The group is diverse in terms of responsibilities and activities; however, for the purpose of our exercise we assume that they should not require sophisticated intermediary dialogue devices between them and the computer. Of course, there must be some form of priority system whereby junior or untrained staff cannot gain access to sensitive material (e.g. access to details concerning salary).

Clerical staff This group is not included under the general heading of computer staff because they will usually be involved with the inputting of data from specific sections of the department and therefore are not under the direct control of the computer unit. It is the responsibility of this group to enter data into certain systems such as the employee database, the heavy equipment inventory, the accounts system, soils lab data, word-processing, databases, and spreadsheets.

The reason why we chose in our example to delegate the work to discrete sections of the department is threefold. Firstly, it conforms to the overall methodology of user-driven systems, which we outlined in Chapters 1–4. Secondly, the sections mentioned already run manual versions of these systems. Thirdly, the data in these bases may be sensitive and should not be widely known outside the confines of the section itself, senior management, and computer management sections.

In your examples you may decide that there is good reason to keep all data input and edit operations initially under the control of a central data processing unit. This may be appropriate in the early stages of implementation specifically if the system is quite new and does not build upon existing manual practices.

If you do want to do this you might be overriden by offended section managers who see their prestige as threatened by the removal of certain even quite mundane activities to the new computer unit.

Clerical operators will require facilities from the computer, such as:

1. Menu driven.
2. Access via code number.
3. Summary screens of certain limited types of data.
4. Help facility.
5. Urgent request facility.

This may look similar to that which we prescribed for senior management; however, this will not take the same form. The level of help will be of the same type but the clerk is only supposed to be inputting data intially, and possibly editing selected data later. No other access to the files is provided. The urgent request facility will operate to allow the clerk to forward notification of potential errors to the computer manager.

8.5 DESIGNING FOR SOCIAL ISSUES

We have already identified the probability that an incoming computer system will have a range of impacts on staff moral, job mobility, and staff seniority. The system you are planning may lead to massive social upheaval in your organization.

Not all change can be known in advance but a range of measures should be set out in advance to ease the implementation of the new system. There are many issues which we may need to confront but here we will focus on de-skilling and under-skilling (see Fig. 8.9).

1. De-skilling—the capacity of computers to reduce the expertise of an individual.
2. Under-skilling—the capacity of an organization to fail to recognize the importance of training existing staff in the use of computers.

As we see in Fig. 8.9, the two forces can be seen as operating on linked axes. For each section of an organization that is adopting a new information system, the planned system should be evaluated in terms of:

- The likely impacts on existing skilled staff in terms of de-skilling.
- The need of the new system for existing staff to be trained in new areas.

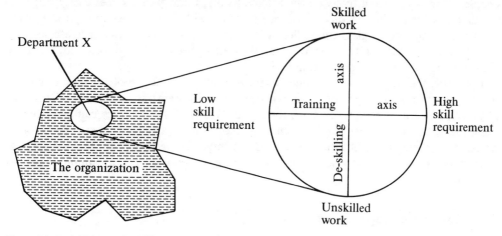

Figure 8.9 Social issues—de-skilling and underskilling.

Ideally any de-skilling tendency should be countered by those staff being retrained in the new technology. This is not always possible.

8.5.1 The case example: specific issues

In our case the impact of the new system was generally seen as being positive. In the short term a certain amount of staff displacement would be encountered but this would run to no more than approximately 4 per cent of the total workforce. The perceived wisdom from meetings with stakeholders in the new system was that this level of disturbance was well within the tolerance level of the organization. Measures to be undertaken might be as follows:

1. Maximizing pre-installation briefings to provide workers with as much detail concerning the new system as possible prior to implementation.
2. Detailed and long-term training inputs for all levels of staff. A core staff to be identified which would be the focus of the new system. This core would be provided with relevant training and longer-term outlines of potential future training.
3. A degree of sensitivity can be agreed with senior managers concerning the displacement of existing staff. Sensitivity measures might include:
   (a) offers where possible to work with the new system;
   (b) reallocation to other, similar departments;
   (c) 'generous' redundancy.

8.6 DESIGNING FOR SECURITY

We will look in some depth at security measures for the system as a whole in the next chapter. For now we will focus on the essential work security of discreet user groups.

The focus for the interface is to balance usability with security. As Fig. 8.10 opposite shows in the form of a glass tube filled with fluid, it is very difficult to increase security without decreasing usability, and vice versa. As systems become better protected, generally they become harder for users to gain access to.

Therefore, the security that any one system requires is a function of the level of hazard in the operational environment as assessed by you. If systems are locked in offices with physical access restricted to key staff, then security within the interface can probably be quite slight. If, on the other hand, systems are in open office space where anyone can gain access, then quite stringent interface procedures might be needed to protect the systems from casual access and/or abuse. 'Stringent interface procedures' generally imply password systems.

8.6.1 The case example: specific issues

In our case study we required a certain amount of security, though it was not thought

Security   Usability

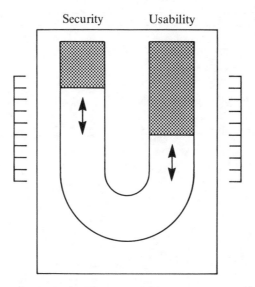

Figure 8.10 Trade off between usability and security.

necessary physically to lock computer systems away. When a manager first enters the system, the computer might enquire as shown in Fig. 8.11 *this screen might protect, but it might also deter. Security is thus a compromise between safety and use.* Users will get no further unless their names and codes match up with those that the computer is expecting. If this match is not made, the potential user might see something like Fig. 8.12 on page 143 —even if there is no monitoring procedure available at all! Often a severe message will put off most of the unwary. For example, most microcomputer systems can easily be adapted to use basic menu systems. Also, many basic database packages come with menu systems available.

The code will also indicate the type of user (e.g. manager, technician, or clerk). The summary screens of information will appear at the end of the process of working through the layers of menus. For example, the senior manager may wish to see if a certain road project is running to schedule. Obviously he or she would select the roads register in the first menu; within the roads register another menu would give details of the services available, one of which might be:

INSPECT PRESENT PROGRESS?

Having selected this, the computer would request details of the particular road—by region, district, and project—and could then give the manager the most recent project report.

As a final note, the *urgent request facility* mentioned for managers refers to a method for diffusing potential frustration as well as providing the manger with a direct means of communicating, via the computer-based information system, with the people who run the computer. With this facility the manager can be provided with means whereby he or she can

```
WELCOME TO THE DEPARTMENT MIS

Please enter your name: _____

Please enter your code: _____




        UNAUTHORIZED ACCESS IS ILLEGAL
```

Figure 8.11

enquire about certain details that may be causing confusion and require further explanation. This could easily be expanded on most available computer networks to allow the user to send messages to other users of the system as well as to the computer staff.

8.7 IMPLEMENTING THE DESIGN

It is not usually the reponsibility of the analyst to create databases and become involved in the processes of database **debugging**, though you may well find yourself in this position. All we can suggest here is that if you find yourself dealing with database set-up/installation as well as analysis and design, you should see the database subsystem in the next chapter; see the references specific to databases in the 'further reading' section of Chapter 9; and follow all database manuals closely in terms of set-up procedures.

The analyst's brief normally runs to suggesting likely software packages which can carry out the types of operation as outlined in Chapters 3 and 4. In situations where computers are being brought into an environment for the first time there is often a degree of flexibility in the adoption of the results of the analysis, and often systems are designed at the point of contact in direct response to the user's wish.

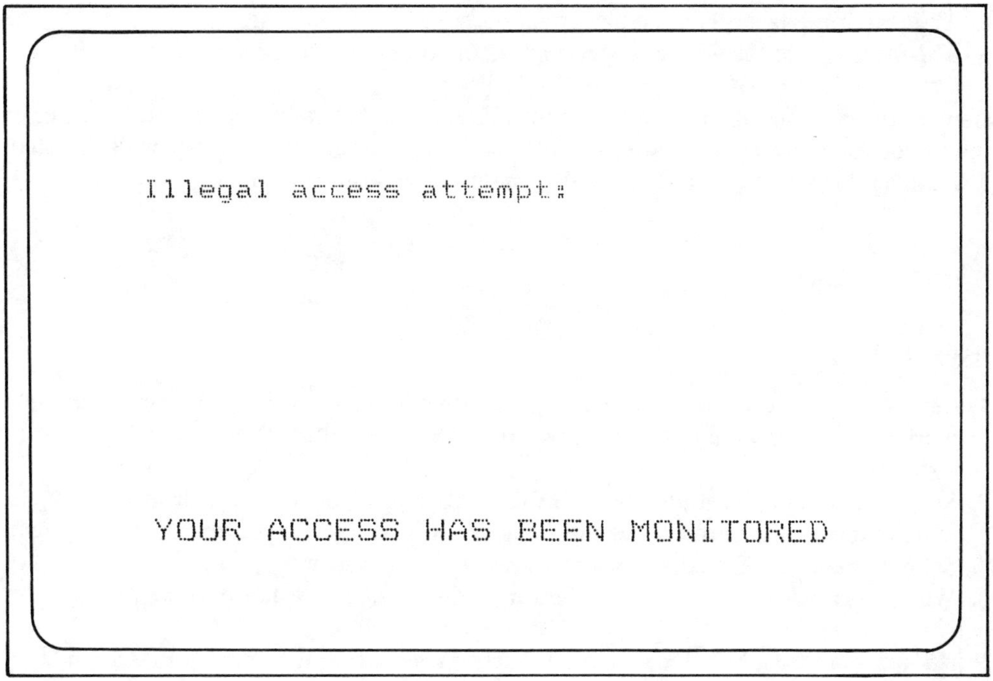

```
        Illegal access attempt:




        YOUR ACCESS HAS BEEN MONITORED
```

Figure 8.12

Amount of time devoted to analysis so far:
Total for this stage (human computer) = 5 days

Cumulative total for path 1 = 25 days
Cumulative total for path 2 = 20 days
Cumulative total for path 3 = 13 days

The interface can take quite some time to get right for all users but we feel that the major principles for interface design in the problem context can be outlined in five days.

8.8 CONCLUSIONS

The interface as set out here has three major aspects:

1. The technical interface—what is seen and how information is put in.

2. The social interface—how the organization copes with new systems.
3. The security interface—how systems are kept secure and at the same time used!

As with the previous chapters, all recommendations in these areas are by way of rules of thumb for incoming systems. However, if these three areas are planned for the final system has a much higher chance of success than might otherwise be the case.

8.9 TUTORIAL

Exercise 5

1. Set out the range of impacts that your information system will have on the existing working relations within the company. Who will be de-skilled? Where may there be staff conflict?
2. Set out a series of measures that you feel will satisfactorily deal with these conflicts.
3. Set out the security procedures that you feel will be appropriate for the new system.
4. Set out examples of ideal screen interfaces for your system.
5. What measures would you plan to ensure reducing staff resistance to the new system?

When appropriate, take a look at the model answer given in Section 9 in Appendix 3.

FURTHER READING

Avison, D. E. and Wood-Harper, A. T. (1990) *Multiview: An Exploration in Systems Development*, Blackwell Scientific, Oxford.

Avison, D. E. and Wood-Harper, A. T. (1991) Information systems development research: an exploration of ideas in practice. *Computer Journal*, **34**, no.2.

Booth, P. (1989) *An Introduction to Human—Computer Interaction*, Erlbaum, Hillsdale, NJ.

Kozar, K. A. (1989) *Humanized Information Systems Analysis and Design: People Building Systems for People*, McGraw-Hill, New York.

Willcocks, L. and Mason, D. (1987) *Computerizing Work*, Paradigm, London.

9

TECHNICAL ASPECTS—WHAT IS NEEDED?

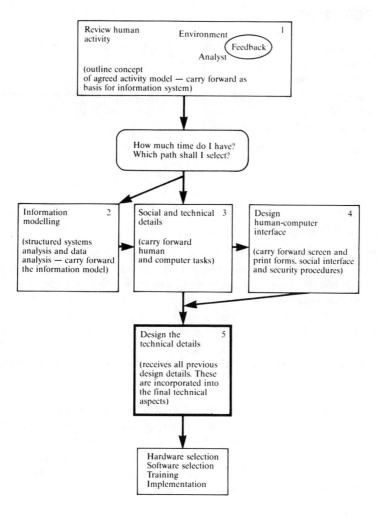

Figure 9.1 Rapid planning methodology.

Keywords technical aspects, applications, database, retrieval, maintenance, management, monitoring and evaluation (M&E).

Summary At this stage of the analysis we begin to design the various component parts that will make up the final information system. The components we deal with here concern the way information activities are coped with by the system, the database structure that contains information items, the production of reports and other output, the management and maintenance of the system, and finally monitoring and evaluation.

9.1 INTRODUCTION

So far you will have undertaken one of three paths to arrive at the techical design stage. As an *aide-mémoire*, the three paths are shown in Fig. 9.2. No matter which of these you have adopted you will arrive now at the stage of technical design (there is no escape!).

Technical design refers to the stage that is concerned with outlining and then combining key components of the information systems which can be usefully planned independently. The six areas that we will focus on here are:

- Applications.
- Database.
- Retrieval.
- Management.
- Maintenance.
- Monitoring and evaluation.

The components link together to produce a 'technically workable system' as seen on page 148 in Fig. 9.3, which shows that this stage is ideally based upon the results of previous stages, e.g. the application area is based upon events identified in Chapter 6, information modelling, the socio-technical system identified in Chapter 7 and the human–computer interface worked through in the previous chapter. Even if you have taken the third, minimal path through the analysis and design procedure, you will be able to make effective use of the rules of thumb that we apply in this stage of the design process.

Before we look at each of the items in detail we can begin by discussing what each of the six areas actually involves.

Application The application is the core of the system you are implementing, e.g. management information system, decision support system, geographic information system, payroll system, accounting system. The application is the 'to do' part of the system; it is all the functions that we wanted carried out on entities (if you undertook stage 2, information modelling). Generally this means that applications refer to activity within the information system to provide the information products desired.

Database The database is at the centre of the new information system. All information

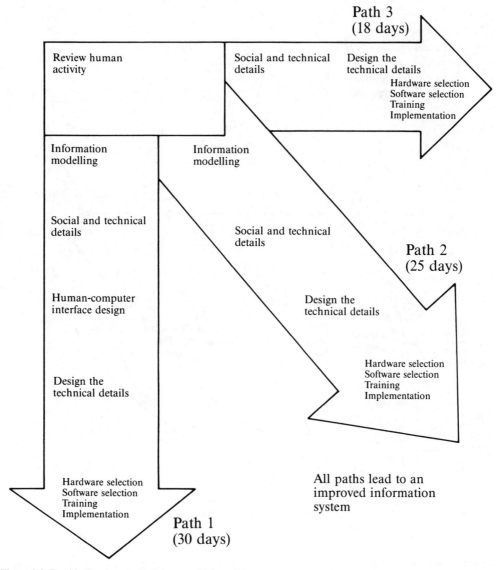

Figure 9.2 Rapid planning methodology—which path?

systems are centred on a database. If you have taken paths 1 or 2 through this book you will have structured the basics of the database already and you will be familiar with the ideas of 'entity' and 'attribute'. If you have taken path 3, then this will be a new idea. The database for your system will be the thing that stores information which your application area will work on to produce the key information products (e.g. reports) required.

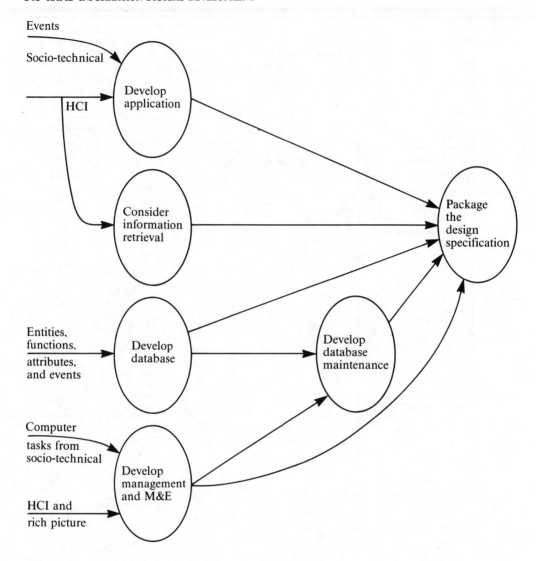

Figure 9.3 Events.

Retrievals The retrieval aspect is the component of the system that produces the information products, be they maps, reports, performance indicators, or whatever.

Management Management runs the overall system. It deals with security, backing up, and all stages of activity within the system. Management is usually in overall control of both of the following areas.

Maintenance This keeps the system going. Under the general title of maintenance we include the planning of preventative and corrective maintenance procedures.

Monitoring and evaluation An aspect that is central and often ignored, monitoring and evaluation ensures that a working system stays working. Monitoring and evaluation, or M&E, provides both management and maintenance subsystems with vital information with regard to the overall health of the system.

Figure 9.4 overleaf sets out one hypothetical combination of these aspects.

Having introduced the area, we will now go into the details of planning each of the six aspects and linking them in one overall schema ready for implementation.

9.2 THE APPLICATION AREA

This element is concerned with the development of what the information system has to do; these are the 'transactions within the system'. One example of the outworking of this part of the process comes from Chapter 6. In the information-modelling stage we suggested that activity could be set out, even at that early stage, as a computer programme. If you are not familiar with this, Fig. 9.5 (page 151) will either serve to remind you or introduce the concept.

Figure 9.5 shows the manner in which four performance indicators (PIs) are produced by a computer package. The package does not interest us at this stage, nor is it important to understand what is going on in detail. The important feauture to recognize is that the PIs are the key information products that we want. Figure 9.5 shows what is required to generate each of these PIs and what information databases (or 'tables') need to be open to allow the applications to work effectively. PI1 produces information on full-time lecturers (FTL), PI2 on total allocatable lecturer contact hours, PI3 on lecturer contact hours, and PI4 on full-time equivalent students.

If you have undertaken the work effectively, a programmer could take your application outline and write a computer programme that could do the job.

The main skill to develop for this phase is to think through the applications sequentially and hierarchically. For example:

How many applications do I have and what shall I call them?
1. Application A—product listing.
2. Application B—relative product value in terms of return.
3. Application C—best performers in the last x months.

What does each application involve? Application A:

1. Open product database.
2. Index the database on product name.

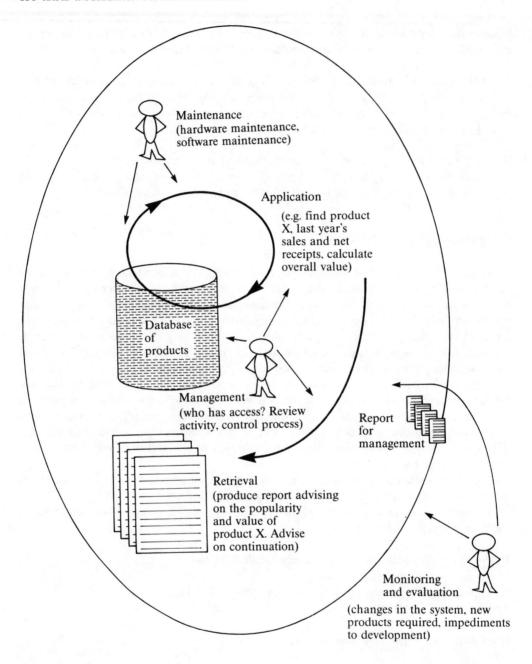

Figure 9.4 Technical areas.

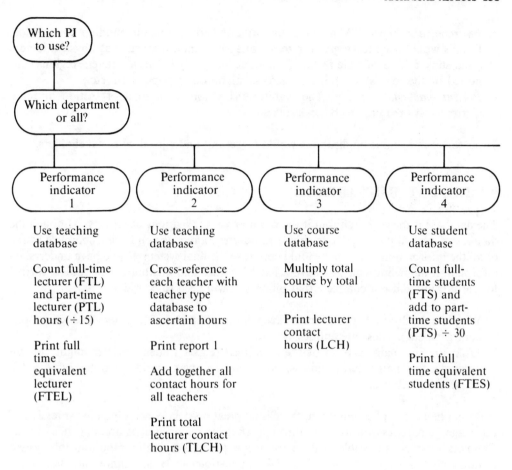

Figure 9.5 Performance indicator (PI) generation.

3. Produce listing of products including present gross return for financial year.
4. Print out listing.

9.2.1 The case study

For our department of roads the application of key concern was the roads register. For this register the particular applications outlined were:

1. *Present cost status.* This referred to the running cost of a particular section of road. The figures would need to be comparable between different regions.
2. *Amount of use.* The physical amount of traffic that a specific road has to cope with.

3. *Environmental impact.* What effect has road x had on the surrounding countryside. Details would need to be given in terms of specific indicators such as physical factors (land slips, etc.), economic factors (economic inequality between different regions connected by the road(s), etc.), social factors (migration of population, etc.).
4. *Present maintenance status.* The maintenance record of a road or section of road. Current tasks required to be undertaken.

Tasks for each of these applications would be set out and agreed with stakeholders.

9.3 THE DATABASE AREA

The database is the core. Without it there can be no applications or reports. Although the database is central it can be quite simple to outline so long as you have a complete listing of all the information items that should comprise the final system. If you have undertaken information modelling you will have a list of entities and attributes that can be carried forward; if you did not complete that phase you will need to consider:

1. What are the things that you want to keep information about? (registers, listings, people, contracts, etc.). These are entities.
2. What are the main features of these entities? (e.g. a student register might contain surname, first name, term address, home address, past marks, disciplinary records). These are attributes.

Rather like the applications area, there is no great mystery concerning the expression of this stage, it requires you to think through what the major entities are and then give as complete a listing as possible of the attributes required, keeping in mind that this system is intimately linked to the applications. The application subsystem cannot undertake work if key attributes of key entities in the database subsystem are missing.

9.3.1 The case study

Continuing the theme from the applications subsystem, the major entity was agreed as being the roads register, and the working out of attributes was as follows:

Attributes for the roads register

Name of road.
Date of construction.
Personnel involved—engineers, overseers.
Duration.
Cost.

Benefit as projected in original report.
Source of finance—external, internal.
Total quantities and costs of:
   cutting,
   filling,
   gravelling,
   culverts,
   bridges.
Maintenance costs.
Total cost to date.
Current conditions.
Secondary developments along road (shops, camps, etc.).
Evidence of environmental degradation:
   land slip,
   pollution.
And so on.

9.4 THE RETRIEVAL AREA

The retrieval area provides the designer with the opportunity to:

1. Structure the content of reports and other output.
2. Agree the timing of reports and other output.
3. Prepare the presentation of reports and other output.

There will be key events when certain information products are required:

- End of the financial year.
- Enrolment of new staff.
- Completion of a project.
- Marketing of a new product.

At these events information of certain types will be required. The form and presentation of the information is also important. One of the most powerful functions of modern computer-based information systems is that retrieval can be automatic and fairly effortless if planned for. On the other hand, if a system has not been properly planned the production of reports can be almost as time consuming as undertaking the process manually. The retrieval subsystem should link in closely with the application and database systems but can be planned separately. The planning process requires the following actions:

1. Assess which reports are required. Make allowances for reports required at a given time

(e.g. end of the financial year) and those required at any time (e.g. the current activity of staff in a design agency). List these reports and indicate when they are required.

2. For each report set out key information that should be included. This can comprise a listing. When producing the list keep in mind the applications subsystem. Applications should produce all the workings for these reports, there should never be a report or other retrieval operation without an application. Similarly there should never be an application that does not produce, or combine to produce some form of report.

3. When you are sure that your listing of all retrieval products is complete and that the content of each of these is similarly set out, think about design. There are two golden rules in the design process:
   (a) Clarity and simplicity.
   (b) Recognizability—resemblance to existing reports.

Reports can contain far too much information, and they can be highly unstructured. To maximize the clarity of a report it is advisable to work with an agreed template. A template is a structure that all reports in a given context will obey. For example, a promotions record report taken from a personnel register might be as follows in template form:

— — — — — — — — — — — — — — — — — —

Surname:
First name:
Home address:

Work address:
Grade:
Pay-roll number:
Last three promotions:

— — — — — — — — — — — — — — — — — —

Do you need to know sex and age? Does 'work address' mean 'name of department'? What does 'grade' mean. Does a promotions report need to contain details of the payroll number? Certainly this form is short, but will it tell a manager what he or she needs to know?

The retrieval system should be planned to provide information at a time and in a form that is useful. The obvious caveat to this is the need to produce all designs in co-operation with the major stakeholders who are going to make use of the reports.

9.4.1 The case study

Following discussions with the major users of the applications, it was decided that existing report forms would be copied and reproduced by the system. There was a certain degree of inflexibility concerning reporting. The computer-based information system was seen largely as a way of speeding up the existing process, the range of new options in terms of reporting was largely ignored and/or not seen as being important. This is a situation that occurs quite regularly. A popular response to this is to design reports around the existing documents but to leave provision in the system for change and further development. Most software packages allow the user continually to add reports.

9.5 THE MANAGEMENT AREA

Management covers a wide variety of issues. For our purposes we will focus on:

1. Controlling your system with the operating system.
2. Job priority control—what to do and when (issues of continuous workload and seasonal workload).
3. Security—access, debugging, and back-up.
4. User support.

9.5.1 Controlling your system with the operating system

This section will not teach the use of the operating system. The object is to point at needs for good management and suggest ways to achieve these needs.

Inevitably understanding the operating system is the bottom line for good management of the hardware and software. Yet training in the use and control of operating systems is a component left off many training schedules. This is a point worth noting and carrying forward to your plan of training needs at the time of implementation. The operating system allows the user to:

1. Control the computer itself effectively:
   (a) understand **error messages**;
   (b) **partition** hard disks;
   (c) check disks for errors;
   (d) control output.
2. Handle software properly:
   (a) **install software**;
   (b) **back-up files**;
   (c) **back-up disks**.

Although the operating system can be laborious and difficult (in fact it can often be the source of problems for new users) it is essential if your system is to work effectively. What problems can arise without it?

- Computer error:
  - overloaded **hard disks**;
  - **fragmented files**;
  - **lost clusters**.

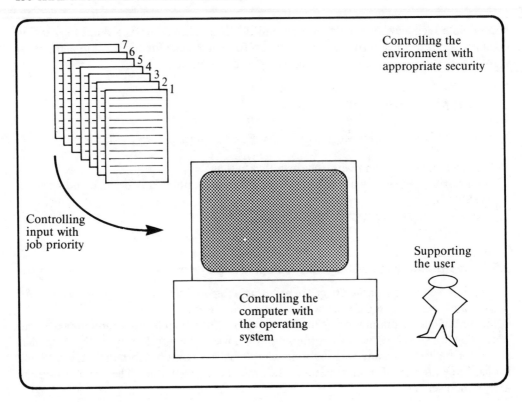

Figure 9.6 The management area.

- Software error:
  - incorrectly installed;
  - incorrect screen/printer driver;
  - **software–hardware mismatch**.
- User error:
  - files lost—no back-up;
  - disks lost—no back up.
- Cost:
  - unnecessary use of potentially expensive outside agencies;
  - lost time;
  - lost work.

Effective training or literature are the major means of avoiding problems. For most purposes a book is good enough if you already have some experience. For the new manager—look to a good training programme to start off.

9.5.2 Job priority control

That is, what to do and when (issues of continuous workload and seasonal workload). Priority is a balance between four factors

1. The job being done.
2. Staff seniority (more senior staff tend to expect quick access).
3. Limited computer time available.
4. Computer frailty.

These four factors can be balanced by the following rule-of-thumb method:

- Arrive at an agreed maximum working day for the machines. A standard applied in many locations is 16 hours per day.
- Review all existing work. Assess for seniority of demand in terms of:
  - seasonal cycles of work;
  - continuous workload;
  - freak loads.

[Seasonal work (e.g. annual accounts) cannot be moved around. There is need to schedule computer availability for key times of the year. Continuous workload—order entering, inventory up-dating—can be slotted in on a more casual basis. Freak loads cannot be planned for. This is the percentage of time that you build in to deal with unexpected eventualities. One way of dealing with this is to make public that the computers are available for 12 hours/day and to hold the extra 4 hours as a buffer against need.]

When you have assessed the workload:

- Produce weekly, monthly, and annual schedules indicating on and off computer times. Most importantly, circulate the schedule with a note indicating the need for discussion and agreement on your plan. The schedule must be known by all users (even if they do not all agree with it).

Once agreed, an annual review is useful, but otherwise—never deviate! It is always a good idea to hold back some computer time in the event of emergencies. This should be time that you do not mind seeing the computer being used but is generally not available for the user.

9.5.3 Security

The most important aspects of security include access, debugging, and back-up.

As we indicated in the last chapter, when dealing with software security at the level of the user interface, as you improve access, so you loose control! Loss of control of the system is to be avoided at all costs. As you increase control, so the user has less and less access!

This is also highly undesirable. Most security therefore involves us in coping with a paradox.

The individual manager sets his or her own balance. Three major aspects of security are hardware safety, software safety, and user safety.

Hardware safety Computer rooms need to be protected from a variety of enemies including:

Excess in climate Climate is of particular importance. Hot, dry, dusty spells in rooms with no air-conditioning can be fatal to microcomputers. Humidity can be a danger. Basic precautions include the following:

- Computer rooms should have doors that make a good seal. An ante-room is a good idea.
- Air-conditioning in computer rooms removes problems of over-heating. Keeping blinds and windows closed is an even cheaper option.
- For very sensitive equipment ensure that air-conditioning provides a positive air pressure so that air will rush out of the room if a door is opened.
- Carpets on the floor remove dust from the atmosphere. But care needs to be taken when vacuum cleaning the carpet.
- During very humid periods, moisture extractors can be fitted in the room.
- Variability of power supply (only an issue now and then).
- Theft.
- Accidents.

Power supply This refers to irregularities in power supply and its criticalness will depend upon, firstly, your location and, secondly, the sensitivity of your equipment.

Some software is very prone to problems with power supply cuts, e.g. some database software corrupts files if they are open when the power goes down. Obviously you must assess your own risk but there are three levels of protection you can apply:

- Total—an independent generator designed to cut in when the mains supply goes down.
- UPS or uninterruptable power supply—these are battery based and will guarantee you 30 minutes work time after power goes down. Rechargeable batteries are usually used.
- Voltage stabiliser—this will regulate any peaks and troughs in your supply. If you are in doubt, get an electrician in to check on your supply.

Theft Local conditions will again prevail, but unfortunately hardware and software are stealable. Levels of protection include:

- Situation of the computer room on the second floor of the building.
- Electronic devices to detect intruders.
- Bars on windows.

- Security staff.
- Searching of staff on entering and leaving the work-rooms.
- Access to keyholder only.
- Securing all computers and peripherals to desk surfaces.
- Providing user identity disks.

Accidents By their nature these are difficult to plan for. The computer manager should get to know the level of risk in the environment. Some of the favourite items to guard against include:

- Tea and coffee being used to irrigate computer systems!
- Disks being erased.
- Keyboards being dropped on the floor.
- Keyboards being hammered by overenthusiastic typists!
- Computers being left on overnight.

Most of these items are dealt with if the manager obeys some basic systems maintenance disciplines:

1. Ensure that all operational staff know what they are supposed to be doing and are kept up to date with systems development. This means that staff training should be continual. Depending upon flexibility of the system the types of staff needed for effective mainte-nance are:
   - (a) The manager—with responsibility for overseeing the total computer system, review-ing policy, liaising with other managers, and reporting to senior management.
   - (b) Systems analyst—improving and developing the system while maintaining the integrity of the existing system.
   - (c) Programmer—operationalizing analysis, design carrying out maintenance on software.
   - (d) Supervisor/adviser—combines control and support to users.
   - (e) Operator—data input.
2. Ensure that all software is copied and protected prior to release. Copies of software should be periodically examined for corruption.
3. Ensure that all hardware is kept in a good state of repair. Measures to provide for this include monthly checking for keyboard failure, **screen burn**, **drive alignment**, and general trouble-free operation.

Servicing (annual) Be careful in your selection of agency! Many servicing agencies are not all that they seem to be. It is worth getting a number of quotations for work and then taking a look at the servicing department of the short-listed companies.

Software safety We have already mentioned keeping the software in a state of physical safety in terms of regular backing-up. Another issue is that of selecting software that offers a clear up-grade path. Often it is difficult to see if software products will still be around in several years time. The best indicator is to go for products that are well supported and have a strong existing customer base. This will be an item that will attract growing attention if software continues to grow as a percentage of the total cost of an information system.

Two other items arise:

- Staff taking liberties. Bringing in their own software and working with it on office machines.
- Sabotage.

Both these items are hard to deal with. Physical protection includes staff vigilance and threatening notices.

Software protection includes the use of lockable keyboards and passwords. Some software does now come with software password control. Packages can also now be purchased; these come with their own password protection, which can be used to keep away the non-expert. But beware: on PCs there is no complete software protection as yet.

Another software threat comes from computer viruses. These are small software packages, written by people who either have nothing better to do with their time or who wish to disrupt computer processes. Viruses can attack any computer system and are transported onto systems (generally) in the form of copied software.

Most recently we have seen the rather tasteless, global AIDS computer virus. This apeared in the form of a disk of information supposedly giving employers information on AIDS in the workplace. Once the software had been loaded onto a hard disk a message appeared threatening to destroy the contents of the hard disk if money was not set to an address for the software antidote. More common viruses are boot sector viruses (e.g. Stoned, Italian, DenZuk) and file viruses (e.g. NME).

Viruses are preventable if you do not copy software. If you get a virus there are quite a number of potential symptoms:

- The temporary interruption of key jobs. This appears to be a common problem. No sooner does an agency or department become dependent upon an information technology assisted system than the system becomes corrupted and ceases to function. The implications are particularly worrying where decisions are delayed in such critical areas as health care, agricultural planning, and product profitability. Close to the root of this problem is the inability of manual practices to cope when the computer fails.
- The closing of information technology departments. Again this seems to be becoming increasingly regular. In one case an information technology department in a university was closed to students because it had just been cleared of viruses for the third time and staff did not want to have to go through it again!
- Unnecessary and expensive hardware replacement. Sometimes hardware is replaced mistakenly when the real fault lies with the software. On several occasions hard disks

have been replaced, and even thrown out when the real fault lay with a virus.

- Slow and or faulty data presentation. Too often computer output is seen as being adequate in itself. Results of calculations are not checked effectively and the garbage that eventually comes out can be used, unknowingly, in management decision making. With the proliferation of management information systems based on computers and the related spread of viruses, decision making is threatened. One of the primary indications of the presence of many boot sector viruses is the slowing down of computer processes.
- Project failure. This is the bottom line. Computer facilities can produce tremendous value; they can also be the weak link that causes projects to fail partially or completely, in the short or long term. Without efficient local data collection, verification, validation, storage, and processing the risk of some form of failure is increased. This is so with manual systems; the risk is even greater when systems are computerized.

What can we do? First and most importantly we need to focus on effective management of information system facilities. There is evidence that at present absolutely minimal attention is being focused on the management of resources. Secondly, culturally and politically sensitive analysis and design should precede all installation. Overly technical approaches to analysis and design produce admirable technical systems. These, however, often break down when put into the social context. Analysis and design needs to understand the context for information systems and recognize that information systems are social systems. Thirdly, planning should be central to pre-computerization and manual systems can often be strengthened and run parallel with incoming information technology based systems. This is a matter of common sense. Any information system requires a degree of contingency planning to ensure that systems can be provided in the event of major breakdown.

User safety We will deal with user support shortly. User safety is improved by:

- Priority access being in place. Users need to know when and how much access they have.
- Passwording in place. A hierarchical structure of passwords can be set up in most systems.
- Ensuring that training is sufficient and maintained. Training schedules for all staff should be outlined.
- Guarding against unauthorized use.

9.5.4 User support

User support has four major themes: access, control, back-up, and feedback. We have already dealt with access.

Control This includes:

- Control over authorized physical access.

- Control over authorized access to sensitive software.
- Control over access to other user's **workspace**.
- Control over user files. If the application that the user is working with is producing data for a third party and that data is critical for decision making (e.g. management information system information), then the user should be guided in appropriate security measures such as always working with copies of files, and regular backing-up of files.

User back-up The manager needs to provide:

- Supervision to hands-on users.
- Training in the use of new packages. This will probably mean that at some time the computer unit will need to provide the users with a curriculum of training activities, including basic introductory programmes, advanced programmes, and programmes in specific applications. Further, training will need to be differentiated between basic operator skills, advanced operator skills, and expert user skills.
- Regular updates on the latest developments. This is a great way of keeping the curious happy, e.g. with a regular newsletter or a weekly/monthly bulletin.

User feedback There is little point in providing users with support if there is no capacity for the user to provide the computer manager with feedback as to the success or otherwise of his or her endeavour. To this end, useful facilities include:

- Regular meetings with the manager.
- A user committee attended by the manager.
- A complaints box.

9.6 THE MAINTENANCE AREA

The lifetime of computer equipment is not a subject that encourages universal agreement. A computer board might advocate seven years for minicomputer hardware; a microcomputer manufacturer might expect 18 months for the desk life of a hard disk. Much will depend upon the wear and tear that the equipment receives and upon the level of maintenance provided. Generally speaking computer maintenance has two major components: *preventative* and *corrective*. Precise details of maintenance procedures can only be given when firm information is available concerning the location of the system and the type of hardware and software to be used. However, in practice we can set out the major requirements of preventative and corrective maintenance as shown in Fig. 9.7.

9.6.1 Preventative maintenance

Taking as our theme the idea that all information systems are set up in situations of hazard,

Example 1. Dangers in the operational
environment

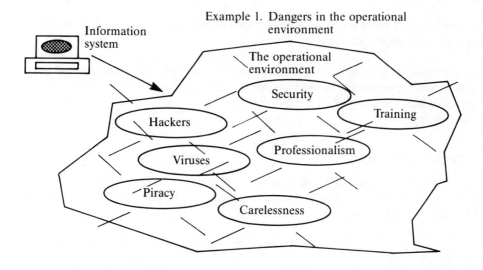

Example 2. Maintenance

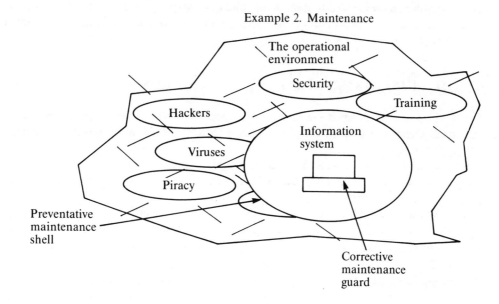

Figure 9.7 Preventative and corrective maintenance.

a preventative 'shell' is essential to provide the system with the bare chance of survival. Features of the shell might include:

1. Daily supervision of the users—this includes the development of appropriate security measures set out in the management subsystem.
2. Regular, quarterly servicing of all hardware.
3. Annual servicing down to computer component level.

9.6.2 Corrective maintenance

Within the preventative shell some unexpected problems will arise. To deal with these problems rapidly and with minimum disruption to users a corrective 'guard' practice is useful. The guard might include the following aspects:

1. In-house understanding of basic fault finding. Operators and supervisors can be trained in fault identification.
2. In-house maintenance technician or a 24-hour call-out procedure agreed with a local maintenance company.
3. Security procedures linked to monitoring and evaluation of key indicators (e.g. wear and tear on machines, incidence of disk theft, incidence of new software appearing on machines).

9.7 THE MONITORING AND EVALUATION AREA

The working system is never static: changes occur. The way to plan for the events is to implement effective monitoring and evaluation (M&E). M&E in the project cycle can be seen in Fig. 9.8.

Monitoring and evaluation is our means for making sure that the system is producing the information we require when we require it, and also that this is occurring at a cost that is proportional to the benefit.

Although M&E is a detailed subject in its own right (see Further Reading), there are at least two levels of rapid M&E that we can plan for in our rapid planning approach, namely rule of thumb and key indicator.

9.7.1 Rule of thumb

Monitoring This comes before and complements the maintenance of the system. Monitoring should be constant and refers to the wear and tear on the computer facility as well as the operational smoothness of its day-to-day running. Monitoring can take various forms:

- Informal periodic review of users by supervisory staff.

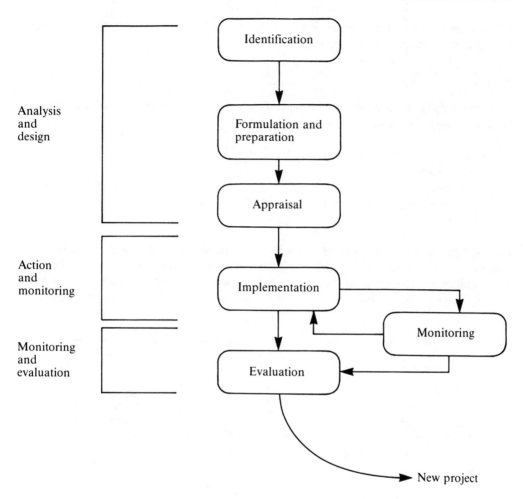

Figure 9.8 Monitoring and evaluation in context. (Adapted from Coleman 1987).

- Formal analysis of user habits, e.g. a questionnaire dealing with:
  - assessment of time spent hands-on;
  - assessment of good and bad habits, such as typing skills, **login/logout** procedure, disk care, and complaints concerning supervision.

All answers need to be carefully interpreted in the light of the obvious user biases.

Evaluation This should be annual and can cover such aspects as:

- The general physical condition of equipment compared to previous year.
- Evidence of theft and/or wilful damage.

- User complaints and/or suggestions.
- Staff complaints and suggestions.

9.7.2 Key indicator

Monitoring and evaluation should deal with all aspects of the information system. This can be seen in broad terms as relating to:

1. The human activity system.
2. The information model.
3. The socio-technical system.
4. The human–computer interface.
5. Technical areas.

For example, if all these areas are regularly monitored for problems and the resulting problems evaluated annually and appropriate action taken, your system should be fairly successful. For the purposes of the exercise here we will not go into great detail concerning the way in which key indicators are identified and monitored. The log frame approach is

Table 9.1 Selection of areas requiring M&E

M&E focus	Critical indicators
1. Human activity systems	New conflicts of interest New departmental linkages Widescale changes of senior personnel Changes in the local economic climate Changes in the local political arena
2. Information modelling	New functions imposed Substantial reworking of old functions Changes in decomposition New entities New attributes New events
3. Socio-technical	Changes in personnel responsibilities Changes in gradings of jobs Economic performance of technology Relative costs of technology
4. Human–computer interface	Changes in operating systems Changes in software Changes in dialogue medium
5. Technical areas	Changes in the system's performance Increasing error levels Retrieval difficulty

powerful and flexible. At this stage it is useful if we have some idea as to the components that are most prone to change. In our own example these were as shown in Table 9.1.

The task of the planner is firstly to assess which of the indicators are to be monitored during the run time of the project and what results would constitute the development of a problem. Secondly, the planner will want to identify several of the indicators that can be evaluated at the end of the project cycle.

9.8 PUTTING IT ALL TOGETHER—THE TECHNICAL PACKAGE

The final package (see Fig. 9.9) can be seen as a specification that will maintain an information system within what is always assumed to be a hazardous operational environment.

The linking together of the aspects into one whole should be occurring throughout the design process.

Amount of time devoted to analysis so far:
Total for this stage (technical aspects) = 5 days

Cumulative total for paths 1 = 30 days
Cumulative total for path 2 = 25 days
Cumulative total for path 3 = 18 days

9.9 CONCLUSIONS

This stage of the analysis has provided you with the outline of the completed system. The only requirements that remain are for the outline hardware, software, and training to be followed by an implementation strategy. Make sure that this is the case! Go through your technical specification and make sure that the system you outline only requires practical details. If your system is deficient at this stage (e.g. if the system does not provide detail for a software package to be described) rectify the point now.

9.10 TUTORIAL

Exercise 6

Using the model from the book, set out, in outline, the overall details of the new system in the form of a briefing document for senior company staff. The document has been

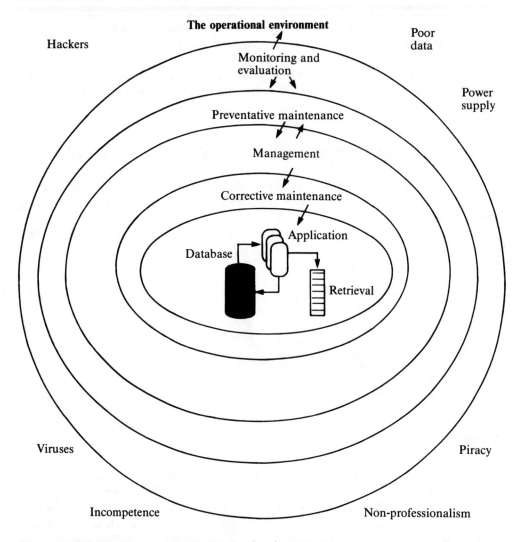

Figure 9.9 The information system in the operational environment.

accessible to a range of different readers and therefore should be quite general. In the document you will need to allude to the total system. This will include items on all the areas outlined above but focusing in particular on:

1. The database and the applications aspects.
2. The management aspect.
3. The M&E aspect. This should include some references to:
   (a) the type of M&E procedures;

(b) expected problem areas;

(c) contingency plans.

Take a look at the model answer in Section 10 of Appendix 3.

FURTHER READING

Avison, D. and Wood-Harper, A. T. (1990) *Multiview: An Exploration in Information Systems Development*, Blackwell Scientific, Oxford.

Casley, D. and Kumar, K. (1988) *The Collection, Analysis and Use of Monitoring and Evaluation Data*, World Bank Publications, Baltimore and London.

Coleman, G. (1987) M&E and the Project Cycle. Mimeo, Overseas Development Group, University of East Anglia, Norwich.

Patton, M. (1980) *Qualitative Evaluation Methods*, Sage, Beverly Hills, Calif.

Rossi, P. and Freeman, H. (1985) *Evaluation: A Systematic Approach*, Sage, Beverley Hills, Calif.

Wood-Harper, A. T. (1989) Comparison of Information Systems Definition Methodologies: An Action Research, Multiview Perspective, Unpublished PhD Thesis.

10

THE TOTAL DESIGN: TRAINING, HARDWARE, SOFTWARE, AND IMPLEMENTATION

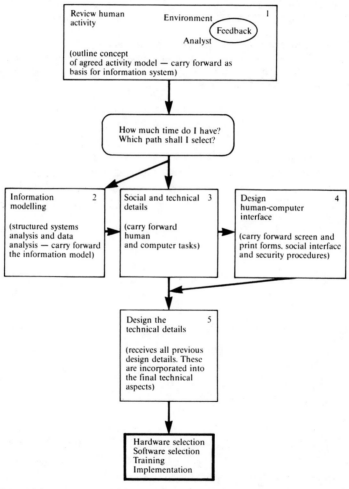

Figure 10.1 Rapid planning methodology.

Keywords training choice, software selection, hardware selection, implementation strategy.

Summary The chapter works on the assumption that you have worked through one of the three paths and that you are now ready to embark on the range of training, hardware and software specification, and implementation issues. Some brief outlines are given of the types of problem that can arise during this phase.

10.1 INTRODUCTION TO IMPLEMENTATION ISSUES

In the previous chapters we have concluded our review of the stages of the rapid multi-perspective approach to planning, which we specify as requiring as little as 30 days (six weeks).

It has been our intention to develop in a concise, stage-by-stage manner the means whereby the methodology can be applied to make sense of the information requirements of the organization on the one hand, and the best-fit solution to meet those requirements on the other.

The total design has been seen to require considerable labour, but we suggest that elements of the methodology can be dropped if time is lacking or certain aspects can be adopted in an abbreviated form. Figure 10.2 overleaf demonstrates the unravelling on the total analysis, design, and implementation model.

Although planning or analysis and design usually ends with the technical specification of the system, implementation issues will quite often be of concern to us. In the next few pages we will look briefly at the four areas of interest for implementation of the design:

1. Training.
2. Software selection.
3. Hardware selection.
4. Implementation.

10.2 TRAINING

What is good training about?

10.2.1 The communication of skills

Information system and particularly computerized information system use is, not as a rule, intuitive. The process of training is usually involved with passing on hands-on skills and confidence in a new system. Other items which may need to be communicated include theory principles, specialist data preparation, and fault finding.

Communication requires that the two sides of the information process are working on

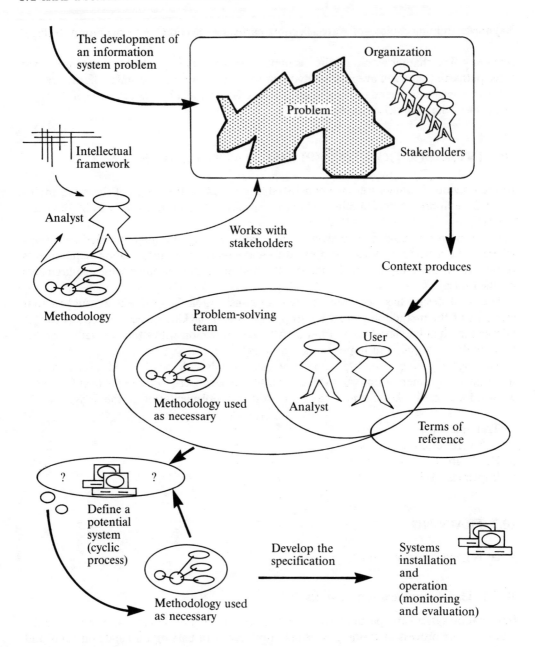

Figure 10.2 The overall approach.

the same wavelength. The training programme that you specify should attempt to enhance communication as follows.

Design factors for effective communication of skills

1. User needs clearly determined.
2. User capacities evaluated.
3. A training pack is planned with a balance of training items and procedures included.
4. A sequence is set out for training.
5. A schedule is designed.
6. **Milestones** of achievement are identified.
7. Final positions are evaluated.

The short cut to effective communication and therefore training is the primary identification of user needs and the secondary evaluation of user capacity. The human activity system and the socio-technical design should have indicated answers to these questions.

The means to effective communication lies in the development of the teaching/training package, the precision of its sequence, and the organization of its schedule. We need to set out the *content*, *sequence*, and *schedule* of events. We deal with this more fully in Sec. 10.2.3.

10.2.2 Repetition of skills out of the training room context—or self-confidence and self-reliance

Training is only effective in so far as the skills developed in the class-room can be replicated at the work place. *This of course is obvious*! However, it is essential that training be developed with this object in view.

Means to achieve self-confidence and self-reliance

1. Including in the training programme plenty of points where the participant has to work on his or her own, e.g. tests.
2. Exercises containing trial runs requiring the use of new skills, e.g. demonstrations.
3. Project work, where the participant can develop their own themes with the existing training material.

The results of not achieving self-sufficiency and self-reliance

1. Folding of projects.
2. Extreme annoyance to the organization running the new information systems as numerous requests for help come in from worried users.
3. Lowering of morale among newly trained staff.
4. Criticism of the training unit that they are not doing their job properly!

Evaluation of performance It is all very well performing the training and assessing that the users are up to the job in-hand; however, one should always expect the unexpected.

Training is only one aspect of computer installation.

- Hardware and software faults can lead to total system collapse or at least the undermining of staff morale.
- Humidity or dust can lead to system faults.
- Poor data collection, validation, or verification all cause problems.
- Trained staff may have a high premium on the local labour market and well-trained staff may leave projects and departments shortly after completing training.

There are also training problems such as:

- New software being used.
- The development of new tasks.
- Large increases in data collection and therefore data processing.

All these points indicate the need for an M&E procedure of the situation some time after the completion of training and the return of individuals to their departments. You should already have these types of contingencies planned for from the previous section.

Such an evaluation might make use of the original rich picture concept set out in Chapter 5 and is best carried out by an individual not directly related with the orginal training but briefed on the training terms of reference and achievements.

10.2.3 Specific training issues

There are a number of easy-to-identify issues that can impede your progress with a training schedule. We will, very briefly, run through these here.

Senior manager intransigence Generally speaking, the more senior the member of staff, the greater will be his or her reluctance to adopt the training schedule. This is further complicated if the manger is powerful enough either to daunt the trainers or just tell them to go away! As we have already discussed in Chapter 2, computers threaten many individuals' views of themselves and of their organizational role. Computerized information systems can offend managers' concepts of what constitutes an effective organization. Also, computers can threaten the employment of individuals.

If a training programme is likely to come up against this type of problem certain factors can be usefully applied:

The key to winning over reluctant participants is empathy. Understanding the origin of reluctance and sympathizing, while at the same time attempting to dispel needless worry, is half the battle.

Rembember, particpants will respond better to training programmes if they are convinced that the training is 'on their side'.

A wide range of abilities If you are using computers it is often the case that computer classes tend to have a wider range of abilities among participants than many other forms of training.

Being 'quick' with a computer is not an indication of intelligence. Rapid response to a computer can demonstrate that the participant is good at reading instructions but is not in fact making any mental effort to *understand* the content.

It is not possible or desirable to have all students working at the same pace. However, order can be imposed on the potential chaos of a class all moving at different paces by:

1. The use of exercises to slow down the faster individuals.
2. Tests to indicate that milestones have been reached.
3. The use of a graduated series of instruction that all can work through at their own pace (note that this severely limits the capacity of lectures to come in at the 'right' time for everyone).
4. Projects that encourage fast and slow participants to work together.

Jargon Jargon puts off the newcomer and confuses everyone. It also indicates that the training is not on the same 'side' as the participant. It sets the training apart and reduces comprehension. On the positive side, the occasional use of jargon can be effective in intimidating aggresive or obstructive individuals. Nevertheless, jargon is generally a negative component. Your training programme will need to create a common language that provides the trainer and the participant with a format in which both can communicate effectively.

10.3 SOFTWARE SELECTION

We have already said that most applications used in information systems design and development (be they management information systems, payroll systems, land assessment systems or otherwise) tend to be a database. We have also mentioned that in terms of the case study used in these notes a range of standard packages will need to be employed, including word-processors, databases, and spreadsheets. What we have not yet covered is a selection procedure (see Fig. 10.3).

There are quite a number of methods for the selection of software. The task is to match the perceived requirements of the users to the capacities of the packages available. The type of information that is needed for the specification to be exact is as follows:

1. *Size.* How large are your datasets going to be? How many records per annum (or quarter, or whatever discreet time unit you are going to use) do you expect to add to the base? How large will the total base be?
2. *Volatility.* This refers to the regularity with which you will be deleting old data items and adding the new ones that will keep the base up to date.

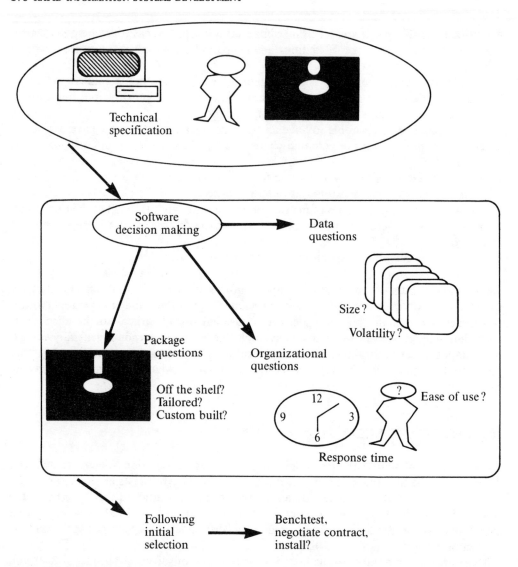

Figure 10.3 Software selection process.

3. *Response time*. At what rate of speed is the data required by the user? Generally speaking, the larger the dataset, the slower will be the response time even if a very powerful computer system is used. There is obviously a need to select the database system that will meet your requirements in terms of speed not just at the present but, using our future analysis in phase 3 of the methodology, on the rate of expansion of the system up to the mid-term (e.g. five years).

With this basic dataset identification information covered, the range of software options available to you needs to be looked into next. In general terms there are three areas of software that might be explored:

1. *Software packages.* These are pre-written and cover an enormous range of applications at the time of writing. What we tend to assume is: 'If I have a software requirement, someone, somewhere has already had that requirement and has produced a package to meet it'. Of course, this may not always be the case, especially in the developmental/ research areas, but it is not a bad working assumption. A review of software catalogues and magazines can identify quite a range, which can then be reviewed more closely.
2. *Tailored packages.* These can be quite expensive and are often required *as well as option 1*. The tailored package is a development of the standard product. It may involve you (or a hired expert) in adapting elements to your specific requirements. Items to be wary of include:
   (a) Poor tailoring leaving considerable **bugs** for you to discover.
   (b) Getting involved in this without thoroughly checking the ready-made market.
   (c) Using spurious experts!
3. *Custom-built packages.* This is undoubtedly the most expensive and most problem-prone area to deal with. All the items to be wary of which we mentioned above still need to be considered. It is a good practice to get any expert group producing such software to give written guarantees and dates for delivery (as well as cost!). Much modern software is produced with quick build tools which provide developers with the ability to significantly reduce development time (Alan and Weiss, 1984).

10.3.1 Case study

For our own example we estimated that any one of a range of software packages would carry out the major database functions that we would be engaged in. The main, specifically *technical* requirements were:

1. For the database system to be **multi-user**.
2. Similarly, spreadsheet and word-processing packages should have multi-user facilities. Specific development factors included:
   (a) packages used must be readily available in local markets (remember, our example is set in a developing country);
   (b) all selections should be well-established products which had the widest chance of being locally serviceable in terms of training.

Realistically we work on the assumption that most software provided for and by developing countries will be pre-written. This will partly be due to the enormous range of software already available and the prohibitive cost of custom-built software. General points to keep in mind (adapted from Kozar, 1989) are the following:

- On reviewing a series of software options, which of the packages actually do the job that you have specified in the logical model of technical specification? Does the package provide you with the levels of control which are essential for successful running?
- Does the software fit the outline hardware configuration that we are forming? Will the software shake hands smoothly with the operating system?
- How well do the performance times of the software rate? Can a trial dataset be run on each system for comparison with standard workloads?
- Can the system be made as user friendly as was set out in the human–computer interface of the methodology?
- How adapatable to change is the software?
- Is the vendor for the package(s) in question honest, legal, and decent; and if not, is this acceptable?

10.4 HARDWARE SELECTION

Having made a choice of software that meets the needs your analysis has arrived at, you are now ready to specify hardware for the system. Again there are a number of key factors that have to be considered (see Fig. 10.4).

The process involved with hardware selection can be delineated as follows:

1. System evaluation.
   (a) Information gathering on potentially appropriate systems:
       - trade magazines;
       - visits.
   (b) Identify vendors:
       - request details of total cost, including all hardware components, hardware training, and hardware support/maintenance.
   (c) Is hardware flexible enough to adapt to the user environment?
   (d) Produce short-list of potential systems.
2. Installation evaluation.
   (a) What precise local factors will affect installation? For example, humidity, temperature, dust, seasonal fluctuation, budget for spares, availability of spares locally, availability of finance.
   (b) How easy is the system to use?
3. Request details of existing user base; discuss the system with existing users:
   (a) Runtime problems in similar environments;
   (b) Beware defensive attitude following 'bad' decision.
4. Test drive (benchmark test) the proposed new system

10.4.1 Case study

For our example we decided upon the microcomputer network. Our final hardware

THE TOTAL DESIGN **179**

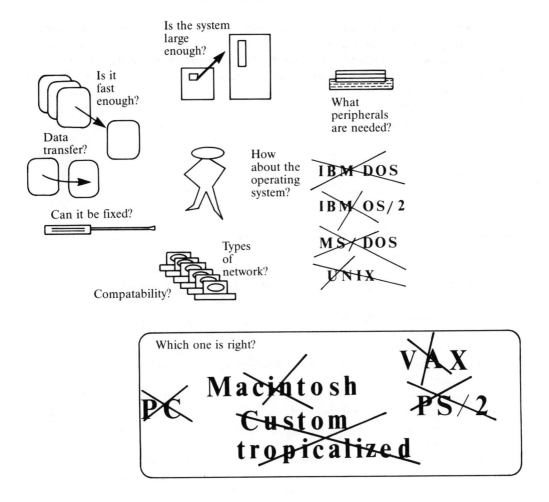

Then: benchtest, negotiate contract, arrange installation

Figure 10.4 Hardware specifications.

selection was for a 80386-based Novell network on PC, XT, AT compatibles with uninterruptible power supplies. The 80386 was chosen because it was the fastest reliable and well-tested chip available at the time. Novell also offered well-established, international support.

In making the decision the major contenders were existing IBM compatibles, or the IBM PS/2. We decided that on the grounds of *local* support and international compatibility with similar activities that the selection should be IBM compatible. We decided against the PS/2 on the grounds of its lack of international availability at the time of the work and also because of the lack of local support, the high cost of spares, and a question mark over its reliability in humid, tropical climes.

Trial MIS operations had already been run on IBM-compatible hardware in similar conditions. The benchmark test was therefore dispensed with.

10.5 INSTALLATION

Following the nuts and bolts of hardware and software purchase, implementation is the next stage. Several factors need to be kept in mind:

1. In new environments rigorous management techniques will need to be applied (see Chapter 9, *the managment aspect*).
2. Staffing will need sufficient initial and recurrent training in the use of the technology (see the *training structure* given in this chapter).
3. The monitoring and evaluation procedure will have to be continued on an annual basis. Initial (within three years) success should not create undue complacency.

Actual systems installation can take a number of forms (see Fig. 10.5). The precise form will depend upon the circumstances in which you are working, and the perceived benefits of each approach as perceived by yourself and the major stakeholders. Kozar (1989) lists four considerations to keep in mind when selecting the installation structure.

Stress considerations The least stressful of the four approaches is probably the parallel system. In this case there is a fall-back and the various agencies within the host institution should not feel too 'exposed' with the new system.

Cost considerations Although the parallel approach appears the most costly, purely in terms of keeping two systems running at the same time, it may reduce some runtime costs because of the reduced stress on operators and managers. This is yet another example of the local environment being the major force in the adoption of a system.

Duration considerations This refers to the speed for system uptake and the length of time it is envisaged that it will take for the system to be partially implemented. Again the parallel approach would appear to be the most costly in terms of duration. Users of the old system are more likely to hang on to functions that they know and are happy with.

Human resource management considerations Resistance is one major factor in terms of staff consideration. The rich picture at the beginning of the analysis should have provided us with some idea of likely centres of resistance. All implementation procedures should be based around a strategy for ensuring minimum disruption to key personnel. The parallel approach should provide the major problems in terms of staff selection for two systems. The

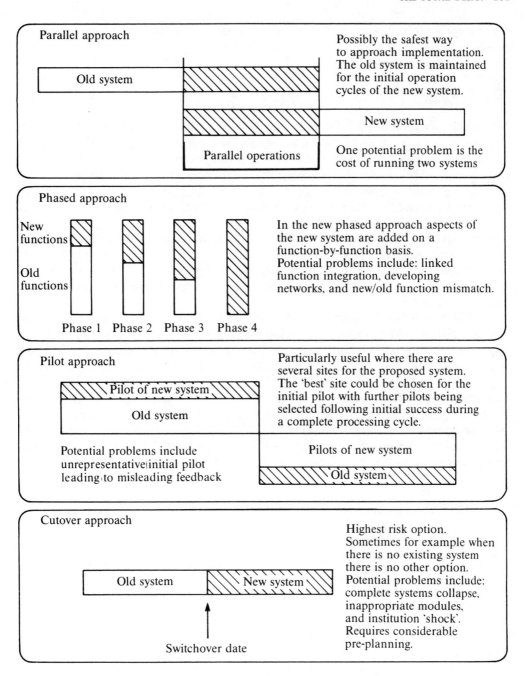

Figure 10.5 Systems installation approaches (Adapted from Kozar, 1989).

cutover approach will produce the most stress in terms of sudden workload and untried procedures.

10.6 CONCLUSIONS

With implementation, the information system moves from the planning and design phase through to operation. You are the proud owner of an information system 'solution', as the technology salesmen like to put it. However, it is dangerous to think of the development process as now being over. Your problems are probably just about to begin! The new information system *will* have teething problems; hopefully your monitoring procedure will pick these up before they become too acute and your maintenance shell and guard should keep the worst of the environmental hazards at bay.

As we have hinted throughout, the greatest concern in the field of information science is undue complacency concerning the power and efficiency of information systems. A healthy attitude to cultivate is that your learning curve is just beginning; the first year or two after installation are often the most valuable in terms of gaining experience—but this period can also be quite painful!

The five stages of the foregoing analysis should provide you with relatively easy-to-apply tool that will allow you to plan and develop information systems for most organizational requirements. Any information system you design will have to be as dynamic as the organization it is going into. If the system cannot adapt to the changing needs of the organization, then you will have problems. Our intention has been to demonstrate not only good practice in information systems planning but to emphasize the need for caution and the capacity to adapt to changing situations. Ultimately the information system is part of the wider social system, and this in turn is part of and dependent upon the global system. The sources of potential interference and disruption are almost endless, but this realization should encourage rather than disincline us to plan as effectively as we can.

FURTHER READING

Alan, M. and Weiss, I. R. (1984) Strategies for End-user Computing: An integrative approach, *Journal of MIS*, Vol 4, No 3, pp 28–49.

Avison, D. E. and Fitzgerald, G. (1988) *Information Systems Development: Methodologies, Techniques and Tools*, Blackwell Scientific Publications, Oxford.

Kozar, K. A. (1989) *Humanized Information Systems Analysis and Design: People Building Systems for People*, McGraw-Hill, New York.

Systems analysis and systems design—some major methodologies in brief

1A.1 'HARDER' SYSTEMS ANALYSIS AND DESIGN METHODOLOGIES

1A.1.1 Structured systems analysis (SSA)

Methodologies based on the structured systems analysis method tend to focus on information movement and analyses. This analysis is broken down in terms of flows, processes, files, sinks, and sources. Possibly one of the best examples we have of this approach is the seven-step model designed by DeMarco and adapted by Martin and McClure (1985). The seven-step model is:

1. Building a current physical model.
2. Building a current logical model from the physical model.
3. Building a logical model of the system to be built consisting of data flow diagrams, a data dictionary, and process specifications.
4. Creating a family of new physical models.
5. Producing cost and schedule estimates for each model.
6. Selecting one model.
7. Packaging the specification.

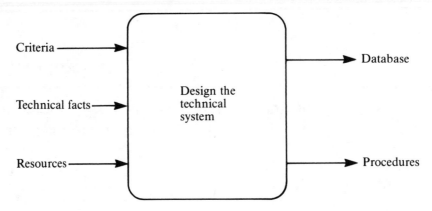

Figure 1A.1 Technical specification.

1A.1.2 Technical specification

In this case the work area consists of a stable data-processing activity. This activity could be mapped and chartered by the use of flow charts and flow diagrams. Formal logic is the main tool used. We can depict the technical specification process is as shown in Fig. 1A.1, where the *criteria* are items such as:

- implementability,
- maintainability,
- flexibility,
- robustness,
- portability (ability to handle various software),
- accuracy,
- security,
- efficiency,
- timeliness,
- compatibility,
- acceptability,
- economy,

(from Waters 1979).

The technical facts concern the interface between man and machine, and the resources are people, money, etc. Figure 1A.2 an example of a technical specification type approach.

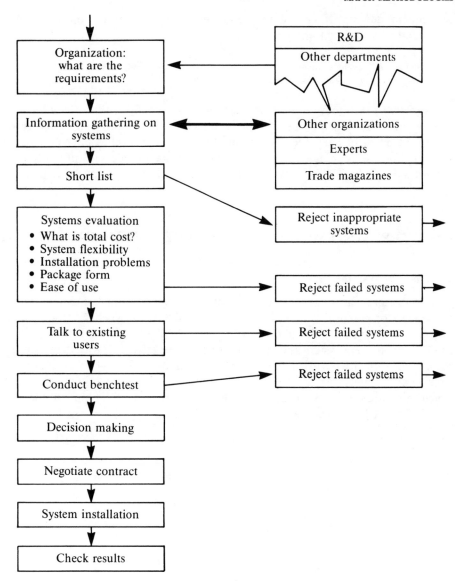

Figure 1A.2 An example that conforms to the technical perspective.

1A.1.3 Data analysis

Here the situation consists of data (Ellis, 1979), therefore if we can identify and map out entities and their associations and properties we will know the full scope of the analysis and the subsequent design. Of course this is a simplification but it points to the root of the methodology. In data analysis

> How a particular system is used can change or be changed, but the underlying nature of the situation remains unchanged because the data are static.   (Wood-Harper, 1989).

It can readily be seen that each of the 'hard' methods briefly outlined here has its own degree of validity, but taken by itself it tends to lack realism in the complexity, risk, and uncertainty of much of daily life. The bonus that hard approaches offer is a very neat methodology for testing things, i.e. *reductionism* (isolating an independent entity); *refutation* (to test an hypothesis); and *repeatability* (any result should be repeatable to ascertain its truth).

1A.2 'SOFTER' SYSTEMS ANALYSIS AND DESIGN METHODOLOGIES

1A.2.1 General systems theory

This is a troublesome theoretic perspective to put into practical application, indeed it is not really intended for practical systems analysis: 'General Systems Theory is too generalised for information systems definition' (Wood-Harper, 1988).

To GST, all reality is a system and therein lies its problem. Dealing with the syncretic and eclectic nature of reality means that you can ignore nothing. Everything is embraced by something else. Change, if it is postulated, has to be vast and utopian in plan. Some have attempted to reduce GST to a working methodology (Orchard, 1972) but still the range is too vast. Table 1A.1 shows a methodology of GST.

Table 1A.1 Methodology of General Systems Theory

1. Isolating phenomenon as an object within an academic discipline
2. Taking a point of view
3. Defining a system consistent with system definition of GST
4. Defining mappings between the object system and the general system
5. Analysing the body of general systems knowledge
6. Collecting the results from the analysis
7. Transforming by mappings to the object system
8. Analysing within disciplines
9. Taking new point of view described and repeat steps 3 and 8

1A.2.2 Human activity system

HAS attempts to operationalize the notion of a human activity system in exploring organizational issues that affect information systems developement. By this we accept that reality is problematic but we recognize within it structure and processes. These twin concepts can be mapped out and understood in terms of a relationship. Checkland (1984) is the main author of this approach and has defined it as 'Soft Systems Methodology'.

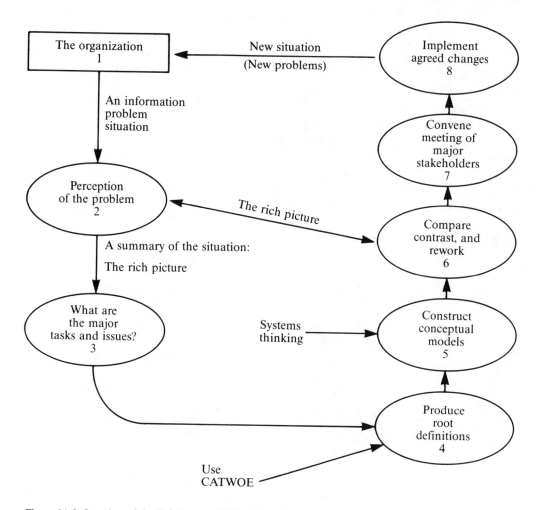

Figure 1A.3 Overview of the Soft Systems Methodology Procedure. (Adapted from Wood-Harper, 1990.)

In this case the problem situation is relatively unstructured. Boundaries are uncertain and these may or may not coincide with organizational boundaries.

Some of the actors in the situation will be obvious, others will materialize later on in the process of analysis. Concerned persons are aware of situations in which improvements are possible. Checkland sets cut seven stages, here we break down to a five-phase structure:

1. *Perceiving*—appreciation of the situation. No imposition of a systems framework.
2. *Exploring*—systems models are constructed (root definitions and conceptual models).
3. *Comparing*—systems models are compared to perceived reality.
4. *Deciding*—what changes are desirable and feasible.
5. *Acting*—implementation of decisions (in some cases including further development of an information system).

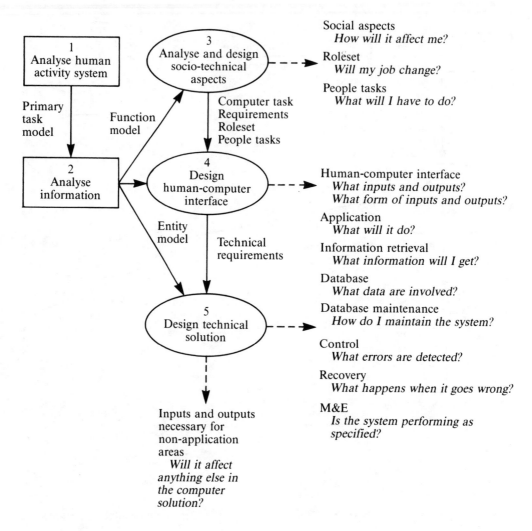

Figure 1A.4 Multiview.

Figure 1A.3 on page 187 shows an overview of soft systems methodology.

1A.2.3 Socio-technical systems (STS)

With STS, information technology and information systems are things that impose new processes on work. Human activity and technology have to be linked into wholeness, with neither dominating or dictating the capacities of the other. In short, there are two sets of objectives: the technical and the social. The primary ideas of the STS approach have been set out by Fok, Kumar and Wood-Harper (1987). The six key headings are:

1. Assumptions about the organization (an open system and a work system).
2. Assumptions about people (participate in decision making).
3. Socio-technical design goals (satisfy organizational and technical requirements).
4. Assumptions about the socio-technical design process (workers should participate).
5. Socio-technical design concepts.
6. The procedural steps of socio-technical design.

The STS approach is more fully developed in Chapter 7.

1A.3 AN ECLECTIC APPROACH—MULTIVIEW

The multiview approach (see Fig. 1A.4 opposite), adapting and adopting elements and tools from the methodologies mentioned here, is the root of the multi-perspective, rapid approach adopted in this book. (See Wood-Harper, 1988 and Avison and Wood-Harper, 1990).

FURTHER READING

Ad Hoc Panel (1987) *Microcomputers and their applications for Developing Countries*, Westview, Boulder, Colo.

Anthony, R. N. (1965) *Planning and Control Systems: A Framework for Analysis*, Harvard University Graduate School of Business Administration, Boston, Mass.

Antill, L. and Wood-Harper, A. T. (1985) *Systems Analysis*, Made Simple Computer Books, Heinemann, London.

Argyris, C. (1985) Making knowledge more relevant to practice: maps for action. In E. Lawler *et al.* (eds), *Doing Research that is Useful for Theory and Practice*, Jossey Bass, San Francisco, Calif.

Ashworth, C. and Goodland, M. (1990) *SSADM: A Practical Approach*, McGraw-Hill, London.

Avison, D. E. and Fitzgerald, G. (1988) *Information Systems Development: Methodologies, Techniques and Tools*, Blackwell Scientific, Oxford.

Avison, D. and Wood-Harper, A. T. (1990) *Multiview: An Exploration in Information Systems Development*, Blackwell Scientific, Oxford.

Balasubrahmanian, A. (1986) Computer scenario in India: a management perspective. Paper prepared for a meeting of the Commonwealth Science Council, 9 October, London.

Bell, S. (1986) Information systems planning and operation in less developed countries (three parts). *Journal of Information Science*, **12**, nos, 5 and 6, pp. 231–245, 319–331, and 333–335.

Bell, S. (1987a) A guide to computing systems evaluation and adoption for users in LDCs—some problems encountered in applying standard techniques. *Journal of Information Technology for Development*, **3**, March.

Bell, S. (1987b) Issues in electronic information: use in the development community. M. Phil Thesis, City University, London.

Bell, S. and Wood-Harper, A. T. (1988a) Information technology adoption and less developed countries (LDCs): towards a best fit appropriate methodology for information systems development. Paper for IFIP WG 8.2 Conference on 'The Impact of Information Systems in Developing Countries', Delhi, December 1988.

Bell, S. (1990) The information technology fix'—lessons from the third world. *Computers in Africa*, forthcoming.

Bell, S. and Shephard, I. (1990) Increasing computerisation in the 'new democracies' of Eastern Europe—lessons from the third world. Paper prepared for a conference in Gdansk.

Bell, S. and Shephard, I. (1990) Information technology and development. Paper prepared for the Development Studies Association Conference, September 1990, Glasgow.

Biggs, S. D. (1982) Agricultural research: a review of social science analysis. School of Development Studies Discussion Paper no. 115, University of East Anglia, Norwich.

Boland, R. (1985) Phenomenology: a referred approach to research in information systems. In Mumford *et al.* (eds) *Research Methods in Information Systems*, pp. 193–201.

Bowers, D. (1988) *From Data to Database*, Van Nostrand Reinhold, London.

Brown, J. and Tagg, C. (1989) A case study approach to teaching systems development. Presented at the 1988 ISTIP Conference Sunningdale.

Checkland, P. B. (1983) *Systems Thinking, Systems Practice*, Wiley, Chichester.

Checkland, P. B. (1984) Systems thinking in management: the development of soft systems methodology and its implications for social science. In Ulrich H. and Probst, G. J. (eds), *Self-organisation and Management of Social Systems*, Springer-Verlag, Berlin.

Checkland, P. B. (1985) From optimism to learning: a development of systems thinking for the 1990s. *Journal of the Operational Research Society*, **36**, no. 9, pp. 757–767.

Checkland, P. B. (1988) Information systems and systems thinking: time to unite? *International Journal of Information Management*, no. 8, 239–248.

Coleman, G. (1987) M&E and the project cycle. Mimeo, Overseas Development Group, University of East Anglia, Norwich.

Damachi, U. G. *et al.* (1987) *Computers and Computer Applications in Developing Countries*, Macmillan, Hong Kong.

DataPro (1980) *How to Buy Software Packages*, DataPro Research

Davies, G. B. (1984) Outline of a talk: challenges in the management of information systems, paper given at the International Conference on Information Systems, Tucson, Ariz.

Davies, G. B. (n.d.) Systems analysis and design: a research strategy macro analysis, Mimeo, Carlson School of Management, University of Minnesota.

Davies, L. J. (1989) Cultural aspects of intervention with soft systems methodology. Ph.D. thesis, Department of Systems, University of Lancaster.

De Marco, T. (1979) *Structured Analysis: System Specifications*, Prentice Hall, Englewood Cliff, NJ.

Douglas, J. D. (1976) *Investigative Social Research*, Sage, Beverly Hills, Calif.

Durham, T. (1988) Steering a middle course between lore and logic, *Computing*, Feb 18.

Eres, B. K. (1981) Transfer of information technology to less developed countries: a systems approach. *Journal of the American Society of Information Science*, **32**, no. 3, pp. 97–102.

Fok, L. M., Kumar, K. and Wood-Harper, A. T. (1987) Methodologies for Socio-Technical Systems (STS) Development: A Comparative Review. 8th International Conference of Information Systems, Pittsburgh.

Gotsch, C. (1985) Application of microcomputers in third world organisations. Food Research Institute, Stanford University Working Paper no. 2.

Grant-Lewis, S. (1987) Computer diffusion in Tanzania and the rise of a professional elite. Paper presented at the African Studies Association Annual Meeting, Denver, Colo.

Han, C. K. and Walsham, G. (1989) Public policy and information systems in government: a mixed level analysis of computerisation. Management Studies Research Paper no. 3/89, Cambridge University Engineering Department.

Han, C. K. and Render, B. (1989) Information systems for development management in developing countries. *Information and Management*, **17**, pp. 95–103.

Haynes, M. (1989) A participative application of soft systems methodology: an action research project concerned with formulating an outline design for a learning centre in ICI chemicals and polymers. M.Sc. thesis, University of Lancaster.

Hirschheim, R. (1984) Information systems epistemology: an historical perspective. In E. Mumford *et al.* (eds), *Research Methods in Information Systems*, North Holland, Amsterdam.

Kaul, M. and Han, C. K. (1988) Information systems for development—the case of Africa. Paper presented at the Development Studies Association of the United Kingdom Annual Conference, September, Birmingham.

Kling, R. (1987) Defining the boundaries of computing across complex organisations. In Boland, R. J. and Hirscheim, R. A. (eds), *Critical Issues in Information Systems Research*, Wiley, New York.

Kling, R. and Scacchi, W. (1982) The web of computing—computing technology as social organisation. *Advances in Computers*, **21**.

Kozar, K. A. (1989) *Humanized Information Systems Analysis and Design: People Building Systems for People*, McGraw-Hill, New York.

Land, F. (1982a) Adapting to changing user requirements. *Information and Management*, **5**, pp. 59–75.

Land, F. (1982b) Notes on participation. *Computer Journal*, **25**, no. 2.

Land, F. (1987) Is an information theory enough? In Avison *et al.* (eds), AFM Exploratory Series no. 16, Armidale NSW, University of New England, pp. 67–76.

Lucas, H. C. (1985) *The Analysis, Design and Implementation of Information Systems*, McGraw-Hill, New York.

Markus, M. L. (1983) Power, politics, and MIS implementation. *Commun. CACM*, **26**, no. 6, pp. 430–444.

Martin, J. and McClure, C. (1985) *Structural Techniques for Computing*, Prentice Hall, Englewood Cliffs NJ.

Mumford, E. (1981) Participative system design: structure and method. *Systems, Objectives, Solutions*, **1**, pp. 5–19.

Mumford, E. *et al.* (1985) *Research Methods in Information Systems*, North Holland, Amsterdam.

Nolan, R. L. (1984) Managing the advanced stages of computer technology: key research issues. In McFarlan, W. (ed), *The Information Systems Research Challenge*, Harvard University Press, Boston, Mass.

Orchard, A. (1972) *The Methodology of General Systems Theory. General Systems Theory Yearbook*.

Pettigrew, A. M (1985) Contextualist research: a natural way to link theory and practice. In E. Lawler *et al.* (eds), *Doing Research that is Useful for Theory and Practice*, Jossey Bass, San Francisco, Calif.

SHELL (1990) *Modelling as Learning, Strategic Planning in SHELL*, Series no. 9, Shell.

Shephard, I. and Bell S. (1988) Application of a systems development methodology in less developed countries. Paper given at the Development Studies Association of the United Kingdom Annual Conference, September, Birmingham.

Symons, V. J. and Walsham, G. (1987) Evaluation of information systems: a social perspective. Management Studies Research Paper no. 1/87, Cambridge University.

Tagg, C. and Brown, J. (1989) TBSD: Notes on an evolving methodology. Computer Science Technical Note, School of Information Sciences, Hatfield Polytechnic, UK.

UNESCO (1984) *Conceptual Framework and Guidelines for establishing GIS*, UNESCO, Paris.

Vickers, G. (1981) Systems analysis: a tool subject or judgement demystified. *Policy Sciences*, **14**, pp. 23–29.

Warmington, A. (1980) Action research: its methods and its applications. *Journal of Applied Systems Analysis*, **7**, pp. 23–39.

Waters, S. J. (1979) *Systems Specification*, National Computer Centre, Manchester.

Willcocks, L. and Mason, D. (1987) *Computerising Work*, Paradigm

Winograd, T. and Flores, F. (1986) *Understanding Computers and Cognition*, Ablex, London.

Wood-Harper, A. T. (1988) Characteristics of information systems definition approaches. Mimeo, School of Information Systems, University of East Anglia, Norwich.

Wood-Harper, A. T. (1989) Comparison of information systems definition methodologies: an action research, Multiview perspective. Ph.D. thesis, School of Information Systems, University of East Anglia, Norwich.

Wood-Harper, A. T. *et al.* (1985) *Information Systems Definition*: The *Multiview approach*, Blackwell Scientific Publications, Oxford.

Wood-Harper, A. T. and Corder, S. W. (1988) Information resource management: towards a web based methodology for end user computing. *Journal of Applied Systems Analysis*, **15**, pp. 83–99.

APPENDIX 2

The project cycle: seeing the task in perspective

2A.1 THE NATURE OF THE CYCLE

The object of this appendix is to set the systems analysis and systems design stage in the perspective of a total project. It is quite easy not to see the fuller picture when you are working on the detail of specific design.

Analysis and design can be said to fall into the stages of the project cycle known as *formulation and preparation*, and *appraisal*. This is not the place to go into great detail on these matters but it is useful to introduce the fuller picture, to set analysis and design in context, and indicate key texts that can be pursued for a fuller appreciation.

Figure 2A.1 indicates the major components of most projects, and shows five distinct phases in the project cycle with the monitoring function as being continuous.

Identification This is usually very general and non-specific. The identification is usually of some general need or general issue such as: 'the need to improve the delivery of sales and purchase data from regional offices to head office in order to give the company a competitive edge', or 'the need to increase the efficiency of core government operations by means of management information system functions'.

The identification stage often occurs without any reference to systems analysis and systems design.

Formulation and preparation This stage is also known as feasibility. It is at this stage that

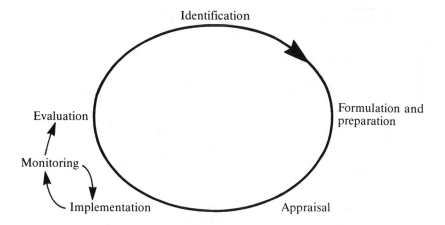

Figure 2A.1 Simplified project cycle. (Adapted from Coleman 1987.)

the workability of and outline terms of reference for the project are decided. If those involved in the formulation and preparation stage think that the project will deliver the required end within the scheduled time, then the real work of planning can begin. Sometimes analysis and design is involved at this time. Quite often an IT-oriented project will make use of a systems analyst expert witness to provide details of the feasibility and cost of given project plans.

Appraisal This is the major planning stage of the project. Appraisal requires systems analysis and systems design because it will be the appraisal document that will form the basis of the implementation plan. It will be during the appraisal period that we would expect our five stages of analysis and design to be undertaken.

Implementation In our book we have set implementation outside the analysis and design stage but still within the analyst/planners mandate of activity. This is often not the case for an entire project. There will often be many non-computer functions related to the project. For example, in our case study used throughout this book there was a communications improvement function which was running with the analysis and design of the MIS and yet had no direct linkage to the analyst/planner. Implementation will mean that all the threads of the project are now brought together. Figure 2A.2 shows the range of factors that may have to be integrated.

Evaluation Most projects should end with some form of evaluation, setting out the original themes for the project work and measuring the level of success in achieving these goals.

2A.2 IN PERSPECTIVE

Systems analysis and systems design is usually a component part of any on-going project.

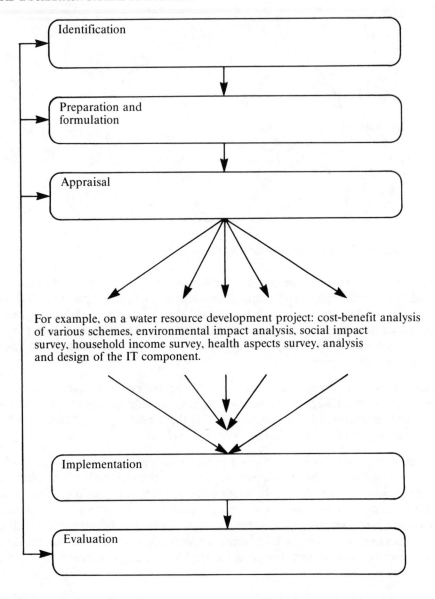

For example, on a water resource development project: cost-benefit analysis of various schemes, environmental impact analysis, social impact survey, household income survey, health aspects survey, analysis and design of the IT component.

Figure 2A.2

Throughout this book we have tried to indicate the importance of the analyst/planner being humble in terms of his or her relationship with those who will inherit the system being devised. Humility is also required with regard to these wider project considerations.

Any work undertaken will need to fit into a greater whole. To some extent all analyst/

planners have to be multi-disciplinary in their approach if they are to survive in the project environment.

FURTHER READING

Casley, D. and Kumar, K. (1988) *The Collection, Analysis and Use of Monitoring and Evaluation Data*, World Bank Publications, Baltimore and London.

Casely, D. and Kumar, K. (1988) *The Collection, Analysis and Use of Monitoring and Evaluation Data*, John Hopkins University Press, Baltimore.

Casely, D. and Kumar, K. (1987) *Project Monitoring and Evaluation in Agriculture*, John Hopkins University Press, Baltimore.

Coleman, G. (1987) M&E and the project cycle. Mimeo, Overseas Development Group, University of East Anglia, Norwich.

Coleman, G. (1987) Logical Framework Approach to the Monitoring and Evaluation of Agricultural and Rural Development Projects, *Project Appraisal*, 2, 4.

APPENDIX 3

A model approach to the exercises

3A.1 THE RICH PICTURE

These are not intended to be pure examples of the techniques discussed in the text. They represent the work of non-expert practitioners using the ideas in the field.

There are several major points to note in the rich picture (see Fig. 3A.1). This is a picture for the analysts use only. It indicates a decision-making spine based on the strong family interests running through senior management. The overriding message from the picture is ignorance. We simply do not know enough about the core-management and administration functions of the organization. The worst mistake we could make at this point is to disguise or ignore this ignorance. We need to know more about core functions. We have read between the lines and indicated a fair degree of frustration and anger in the management section. Initial indicators would suggest that more details are required concerning:

1. Present use of computing.
2. Role and attitude of administration.
3. The working relationship between the three departments.
4. Attitude of the regional offices to head office practices. This is the result of a sideways look at the project. The project terms of reference are stressing MIS at headquarters level. The business of the company takes place in the regions. Are the terms of reference right?

These items could be further developed in a second rich picture.

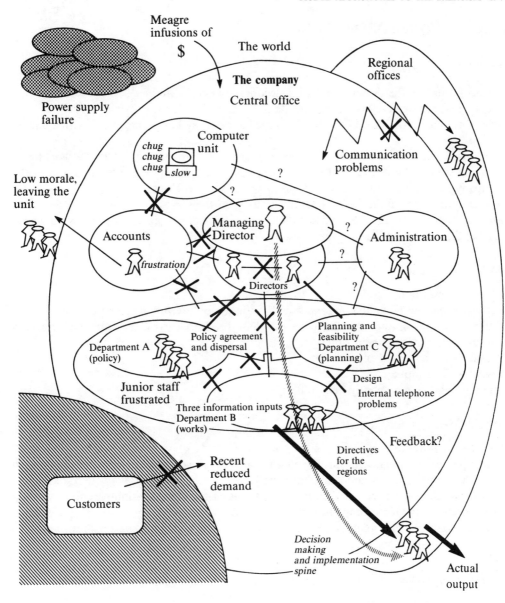

Figure 3A.1 The rich picture.

3A.2 THE ROOT DEFINITIONS

3A.2.1 The analyst

Client—primarily the donor (the bank), secondly the company.
Actor—self, senior and MIS company staff.
Transformation—MIS for key functions aimed at assisting key management with a unified purpose.
World view—improved efficiency of operations and the reduction in internal wrangling. A sneaking feeling that the main problem is not so much the management function of the company but its line function, i.e. the dirty work in the regions.
Owner—the company in the long term, the bank in the short.
Environment—company head office initially, then the regional offices. All work would need to take the regional offices into account.

3A.2.2 The donor

Client—primarily the donor.
Actor—primarily the analyst working within the donor's terms of reference.
Transformation—improving the efficiency and profitability of the company through improving the management function.
World view—increased profitability.
Owner—the company.
Environment—the company head office.

3A.2.3 Head of department

Client—primarily the company.
Actor—company employees working with the analyst. Responsibility for work undertaken is seen as being on the shoulders of the analyst.
Transformation—efficiency, MIS functions, status of the company improved.
World view—improved competitiveness and status.
Owner—company and analyst.
Environment—the company.

3A.2.4 Consensus

Client—donor and company.
Actor—company employee and analyst working to agreed terms of reference.
Transformation—action to develop efficiency and profitability.
World view—cost effectiveness.
Owner—the company and in the short term the analyst.
Environment—the company.

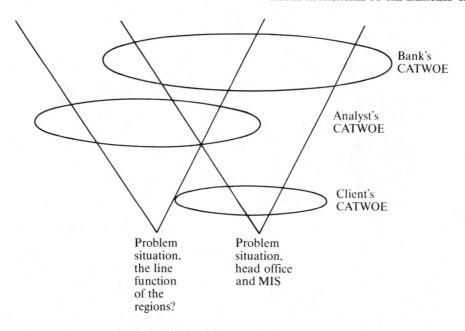

Bank's
CATWOE

Analyst's
CATWOE

Client's
CATWOE

Problem
situation,
the line
function
of the
regions?

Problem
situation,
head office
and MIS

Figure 3A.2 Overlapping root definitions of the problem.

This consensus is not completely agreed. The view of the analyst is not quite that of the bank or the managing director. Putting the view in a cone format, it would look like Fig. 3A.2.

It is not possible for the analyst to alter the terms of reference for the entire project, but it is worth noting down at this stage that there is a divergence of views.

3A.3 THE TOP-LEVEL CONCEPTUAL MODEL

The top-level conceptual model (see Fig. 3A.3 overleaf) does take some liberties with the rich picture. We assume here that there is a clear line of command between the three departments working its way out to the regional offices. We have taken a further liberty by salving our consciences concerning the needs of the regional offices by setting out their needs as a second project. The other point of note is the strategy system lying outside and above the management system. The reason for this is the impact of the bank on strategy.

3.A.4 THE ENTITY MODEL

The entity model (Fig. 3A.4 on page 201) shows that we have simplified the new MIS down to two primary areas—daily administration on the one hand and core planning on the

The company

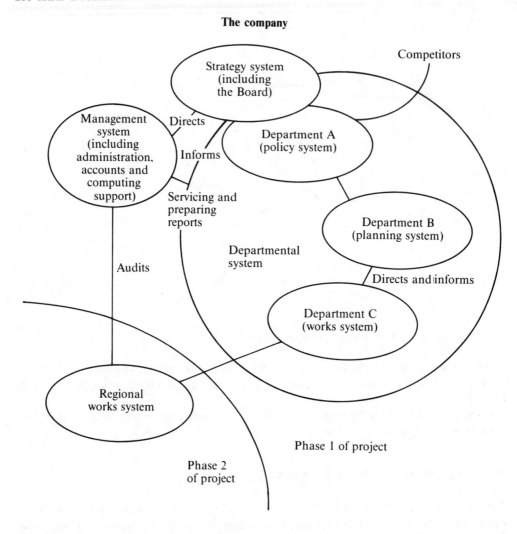

Figure 3A.3

other. Administration has to do mainly with equipment, staff and payroll details; core planning is concerned with project progress. This model could be further simplified—see Fig. 3A.5, page 202—but we thought that this line of approach was too lacking in detail so we opted for Fig. 3A.4.

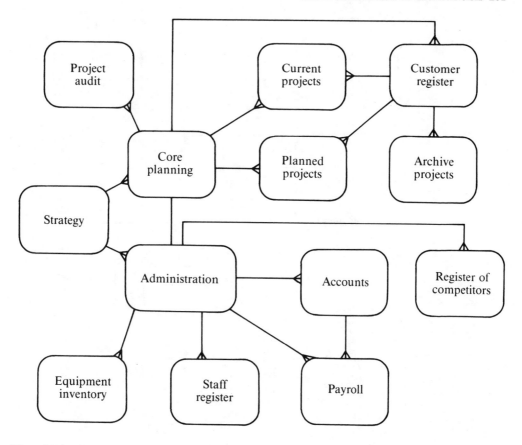

Figure 3A.4

3A.5 PRIMARY ATTRIBUTES

Obviously these details could be developed. The intention here is to show the key attributes only and the attributes which are common to two or more entities (see Fig. 3A.6, page 203). In our example you may well wish to develop more complete attribute lists giving more detail.

3A.6 MAJOR FUNCTIONS

The basis of all functions in the MIS is to support the strategy formulation function. This is therefore our logical first level (see page 204 Fig. 3A.7). Figure 3A.8 on page 205 develops the second-level theme of strategy formulation. To carry the analysis one level further, we look at the details of the review of major competitors (see Fig. 3A.9 page 206).

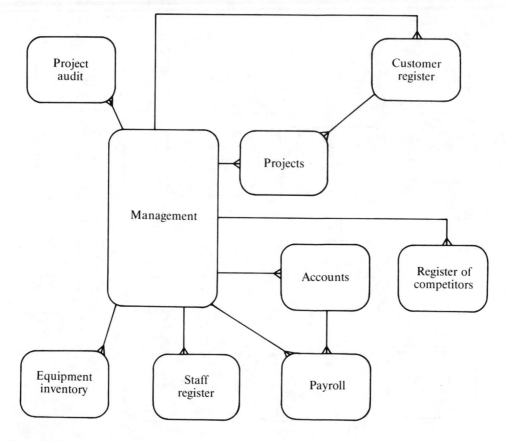

Figure 3A.5

3A.7 KEY EVENTS

The dataflow diagram on page 207 (Fig. 3A.10) shows the action of events upon the functions related to the review of major competitors.

3.A.8 SOCIO-TECHNICAL MODELLING

The approach taken here to socio-technical modelling (see Table 3A.1, page 207) is slightly different to that shown in the main text of the book. Here social objectives are seen to drive technical objectives, i.e. each social objective works its way out in one or more technical objectives.

The alternatives arising from the objectives given are as shown on page 208 in

Core planning

Project value
Future projects
Equipment

Strategy

Accounts
Audit

Management

Projects
Staff

Accounts

Project details
Payroll
Equipment
Materials

Payroll

Staff name
Address
Grade
Salary
Allowance
Tax rate
Expenses

Staff register

Staff name
Address
Record
Grade

Equipment inventory

Name
Age
Location
Value
Depreciation

Project audit

Project accounts

Current projects

Name
Value
Start date
Completion date
Manpower
Materials
Delay
Contingencies

Planned projects

Name
Value
Start date
Completion date
Manpower
Materials
Delay
Contingencies

Archive projects

Name
Value
Start date
Completion date
Manpower
Materials
Delay
Contingencies

Register of competitors

Name
Activity
Value
Contacts
Personnel of
 interest

Customer register

Name
Address
Previous work
Value to date
Scheduled work

Figure 3A.6

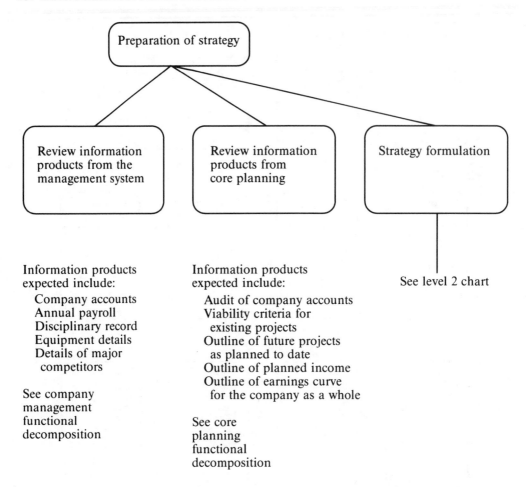

Information products
expected include:

   Company accounts
   Annual payroll
   Disciplinary record
   Equipment details
   Details of major
     competitors

See company
management
functional
decomposition

Information products
expected include:

   Audit of company accounts
   Viability criteria for
     existing projects
   Outline of future projects
     as planned to date
   Outline of planned income
   Outline of earnings curve
     for the company as a whole

See core
planning
functional
decomposition

See level 2 chart

Figure 3A.7 Level 1 functional decomposition: strategy.

Table 3A.2. The main point of note here is that we are linking social and technical alternatives in combination (e.g. T3 = Micro network + T1).

The best two options shown in Table 3A.3, on page 208 indicate a desire to base any new system on new staff bringing in skills at present unobtainable in the company. The point difference between S3T2 and S4T2 is the level of retraining within the company—all or just key staff. In terms of making a proposal based on this analysis it would be quite simple to indicate an expansion path for the incoming system, from a small, highly trained group dealing with the initial microcomputer/manual system, gradually spreading out training to the wider community as appropriate. This would also link to the analyst's indicated preferred direction of the project as a whole—namely outward to the regions.

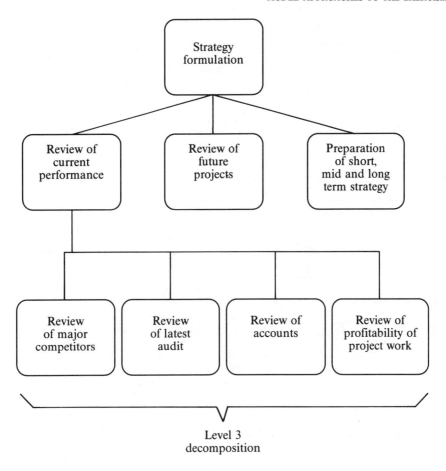

Figure 3A.8 Level 2: strategy formulation.

3A.9 HUMAN—COMPUTER INTERFACE

3A.9.1 Impact of the new information system on the company

Because we have opted for a development path working off established manual practices, and because there is an existing computer unit, we might expect the impact of the system to be fairly minimal. This would certainly not be the case if a new and very powerful computer-based information system were being installed from scratch. The main problem will probably lie with the existing computer staff having to introduce and work equitably with the new, highly trained staff being brought in. Also, as a wide-ranging change in practice will eventually follow the new system, we might expect senior management to need

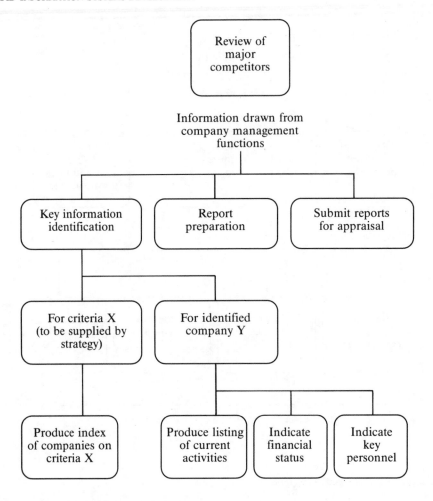

Figure 3A.9 Level 3: review of major competitors.

a considerable degree of awareness training and general encouragement. There may be some displacement of clerical workers but hopefully most can be moved into the new areas of data entry and data edit.

3A.9.2 Suggested measures

1. General firm-wide awareness raising.
2. Training on all new equipment for existing computer staff.
3. General keyboard and computer package use training for clerical staff.
4. Senior management awareness training.

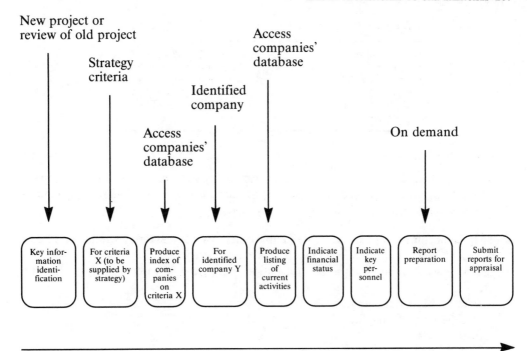

Figure 3A.10 Level 3: events related to review of major competitors (no fixed relationships to date).

Table 3A.1

Social objectives	Technical objectives
Appear credible	Improve communication
	Timely decision making
Improve professionalism	Improve technical skills
Improve budget	Improve timeliness in budget planning
	Demonstrate effective budgeting
	Demonstrate efficient subcontracting
Improve planning	Technical skills
	Automated features
	Networking decision making
Improve office management	Networking
	Standards
	Rapid processing

Table 3A.2

Social alternatives	Technical alternatives
S1. Retrain key staff	T1. Computerizable manual system
S2. Retrain staff	T2. Micro stand alone + T1
S3. New staff + S1	T3. Micro network + T1
S4. New staff + S2	T4. Network

3A.9.3 Avoiding risks

We have avoided the worst risks for a new system by avoiding a networked computer option, although this may occur at some time in the future. A manual/stand-alone microcomputer system can be made fairly secure by ensuring:

1. That all machines are in safe zones (lockable offices with access only to authorized staff).
2. A procedure for authorizing staff to access the system.
3. Levels of access to any computer information. Access would work broadly on the lines of:
   level 1—basic data input,
   level 2—basic data edit,
   level 3—data edit,
   level 4—department-wide information retrieval,
   level 5—company-wide information retrieval,
   level 6—basic system fault finding,

Table 3A.3

Alternative combination	SCsts	TCsts	SCon	TCon	SRes	TRes	Total	Rank
S1T1	2	2	2	7	4	4	21	2
S1T2	2	4	3	6	4	5	24	
S1T3	3	6	4	6	5	5	29	
S1T4	4	7	5	6	6	6	34	
S2T1	3	2	3	7	4	4	23	
S2T2	3	4	3	6	4	5	25	
S2T3	3	6	4	6	5	5	29	
S2T4	4	7	5	6	6	6	34	
S3T1	4	2	3	7	3	3	22	3
S3T2	4	2	3	5	3	3	20	1
S3T3	4	3	4	3	4	4	22	3
S3T4	4	4	4	4	4	4	24	
S4T1	4	2	3	7	3	3	22	3
S4T2		2	3	5	3	3	20	1
S4T3	4	3	4	3	4	4	22	3
S4T4	4	4	4	4	4	4	24	

level 7—access to the total system.
4. Usual software and data back-up procedures as set out in this book.

3A.9.4 Screen interfaces

The priority with a mixed manual/microcomputer system is for the two components to work well together. Standard paper forms will still be in evidence, particularly in the early stages of the operation. We would therefore expect that screen interfaces (in terms of data entry) would imitate these standard forms and be based on a form-filling practice, as shown in Fig. 3A.11 overleaf.

On the other hand, the managers who are going to be using the final information products will require a system that is menu based (see Fig. 3A.12, page 211). It is assumed that computer staff will not require specialized screen interfaces.

3A.9.5 Measures to reduce staff resistance

The key to reducing staff resistance is the demonstration of:

• The value of the system to the individual user.
• The fact that the system does not threaten.

The awareness training sessions are the ideal vehicle for this type of assurance. Awareness raising sessions involving computers too often tend to focus on the use of computers rather than the use of information systems. Ideally a replica of the incoming system should be provided for the users and a range of normal tasks should be set out which will demonstrate to the user that:

1. The system will make life easier.
2. The system is open to change if there are problems.
3. The system should result in a notable increase in the efficiency and work satisfaction of the user.

Where there are problems (widescale changes in work practice, changing job terms of reference) these can be introduced in the context of other positive elements (more pay, better conditions).

3A.10 TECHNICAL ASPECTS

Technical briefing to all senior company staff

The outline of the new information systems The system arising from the review to date can be seen as having at least six distinct component parts:

```
Construction Project Form 1A    Screen 1 of 6

Name of project:    ┌──────────────────────────────────┐
                    └──────────────────────────────────┘
                    ┌──────────────────────────────────┐
                    └──────────────────────────────────┘
                    ┌──────────────────────────────────┐
                    └──────────────────────────────────┘


Location            ┌──────────────────────────────────┐
                    └──────────────────────────────────┘
                    ┌──────────────────────────────────┐
                    └──────────────────────────────────┘

Start date          ┌──────────────────────────────────┐
                    └──────────────────────────────────┘

Proposed            ┌──────────────────────────────────┐
completion date     └──────────────────────────────────┘

Type of project     ┌──────────────────────────────────┐
(new or maint)      └──────────────────────────────────┘


Initial budget      ┌──────────────────────────────────┐
                    └──────────────────────────────────┘

Officer in charge   ┌──────────────────────────────────┐
                    └──────────────────────────────────┘

Team size           ┌──────────────────────────────────┐
                    └──────────────────────────────────┘

Team details (numbers)

Senior managers     ┌──────────────────────────────────┐
                    └──────────────────────────────────┘

Senior engineers    ┌──────────────────────────────────┐
                    └──────────────────────────────────┘

Engineers           ┌──────────────────────────────────┐
                    └──────────────────────────────────┘

Labourers           ┌──────────────────────────────────┐
                    └──────────────────────────────────┘

Details for machinery on screen 2. Hit PgDn for 2.
```

Figure 3A.11

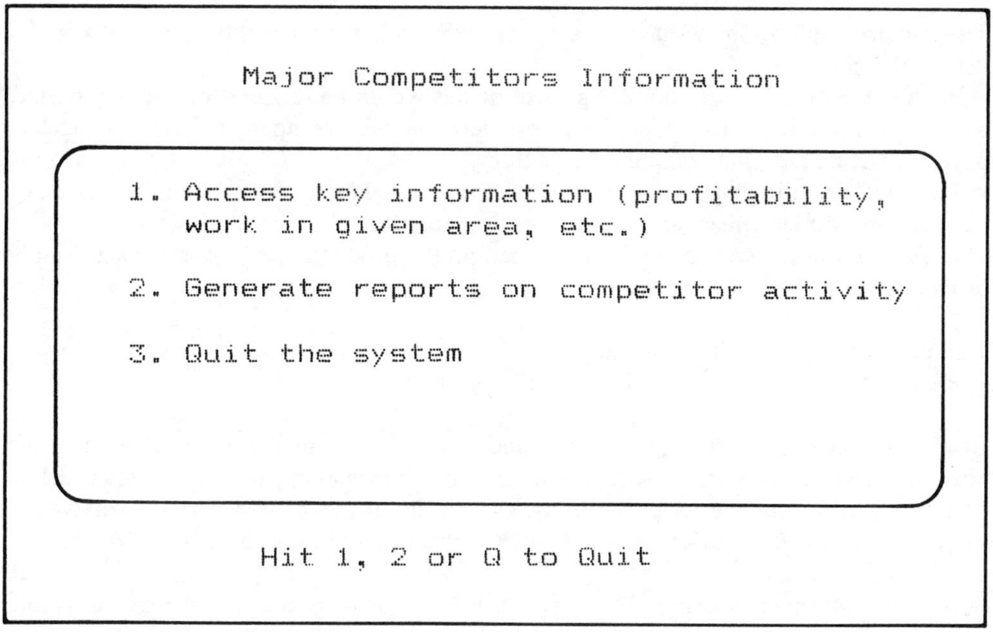

```
            Major Competitors Information


    ┌──────────────────────────────────────────┐
    │                                          │
    │   1. Access key information (profitability,│
    │      work in given area, etc.)           │
    │                                          │
    │   2. Generate reports on competitor activity│
    │                                          │
    │   3. Quit the system                     │
    │                                          │
    │                                          │
    └──────────────────────────────────────────┘


            Hit 1, 2 or Q to Quit
```

Figure 3A.12

1. A use of an application.
2. A database.
3. Retrieval in the form of reports.
4. Maintenance of the system.
5. Management of the system.
6. Monitoring the working of the system and evaluating annual success.

Without going into too much detailed, technical information now, we will set out the general details of the system.

From our analysis and design the overall need for the application is quite clear. Strategy formulation has to be supported by information in the headquarters of the company being made available in a packaged format. The information required for strategy is derived both from the general atomization and also from core planning. Therefore the overall application is designed as a management information system—but more specifically as a strategy support system. The application is designed for interaction at three levels:

1. The production of strategic information products for senior managers.
2. The input of basic data by clerical staff.
3. The control of the system by computer staff.

The primary implications of such a system can be seen in the information to be stored, i.e. the databases.

Initially we would expect that all project details would be required on a computerized database. This information would be cross-referenced with company information such as payroll, staff details, and equipment inventory.

These information areas would be related to and further supported by information on current and past customers and a register of competitors.

The activity of the system would be centred on the production of reports. These can be broken down into two key areas:

- Reports for the support of strategy.
- Reports for the increased efficiency of the company.

Reports in support of strategy These would relate to the major areas of competitive advantage over competitors, the review of current performance in specified areas such as cost and control and project profitability, the results of company audit, and analysis of future project trends—key areas to become involved in, old areas to drop.

Reports for increased efficiency These would need to focus on the effectiveness and value of staff, staff costs as set against returns, the effective use of capital plant, internal accounting, staff records, and staff turnover.

The applications, database, and reporting areas would come under the overall control of management.

The management of the information system In our case management would initially be undertaken by existing computer staff but would eventually probably become the responsibility of new staff to be employed. The central need for management is to ensure that the system is maintained secure irrespective of changes in the company. Effective management requires that the following areas be effectively controlled:

1. The operating system of the incoming system. This includes ensuring that all user interfaces are in place and that levels of control are established through the use of the operating system.
2. Job priority control. Computers cannot provide all information to all users at all times. The central objective of the system is the support of strategy. Access to the system for this priority job would need to be ensured. Strategic decision making is often quite unstructured and requires that the system have the capacity to respond flexibly to need. This may mean that one computer unit is constantly available for strategy enquiries. The information for the management and administration of the company can normally be scheduled in terms of the production of daily, weekly, monthly, and annual reports. Clerical input has to function effectively if the information products are to be provided. Clerical use needs to be carefully controlled and information products need to be passworded off from causal or accidental access.

3. Security, building on the items raised in 2; security for hardware (power supply, theft, and accidents) can be improved if effective control is exerted over the working environment. Because the system will initially be used by only a small number of staff, the threats to hardware should be reduced through control over physical access to the machines.

4. User support. We have already mentioned a wide range of training support for the various user groups. Related to this should be effective day-by-day supervision of users (a function that can be provided by the computer unit staff) and a clearly defined line of communication from the user body back to those who control the system. This is important in allowing users to feel that they have some say in the future development of the information system. A user support committee with attendance by computer unit staff should fulfill this need.

One aspect of effective system management is to ensure that regular maintenance is carried out. The initiation of a preventative maintenance shell linked to corrective maintenance guard is dependent upon the company getting appropriate staff trained in these areas or employing new staff who already have these skills. Given the hazardous nature of the environment in which the company operates (as depicted in the rich picture) the maintenance area will require development prior to system implementation.

Finally we come to the monitoring of the process of the system installation and the evaluation of preliminary information products. Table 3A.4 indicates key items to monitor and the critical indicators to watch.

Table 3A.4 Selection of areas requiring M&E

M&E focus	Critical indicators
1. Human activity system	New conflicts of interest between family members Changes in the local economic climate Changes in the relationship between headquarters and regions
2. Information modelling	New functions arising from strategy Changes in decomposition New entities required New attributes for existing entities New events requiring new reports
3. Socio-technical	Changes in personnel responsibilities Settling in of new staff Performance of technology Integration of manual and computer systems
4. Human–computer interface	New dialogue systems required relating to new information
5. Technical areas	Effective user support? Regular monitoring

The final evaluation of the system should occur following a three-year run cycle to give the new system time to adjust to the environment. The evaluation should focus on the relative performance of the new system as set against the details of the systems analysis and systems design contained here. Of critical importance will be the reporting procedures—are the strategic reports being produced on demand and are these reports of relevance in developing better planning policies?

4GL (fourth-generation language) Computer programme that is semi-intelligent, allowing the computer to be more interactive with the user.

Attributes The basic features evident in an entity.

Back-up files/disks The process of making security copies or back-ups of files and disks of files.

Bugs Faults in computer software.

CATWOE Elements considered in formulating root definitions. The core is expressed in T (transformation of some entity into a changed form of that entity) according to a declared *Weltanschauung*, W. C (clients): victims or beneficiaries of T. A (actors): those who carry out the activities. O (owner): the person or group who could abolish the system. E: (the environmental constraints which the system takes as given).

Computer-based information system The computer is the source of the system which provides on demand a number of key information products requested by an identified user community.

Conceptual model The structured set of activities necessary to realize the root definition and CATWOE.

Consolidated computing The second epoch of the computer age. Mainframe based and programmer orientated.

Context A word always being used by information system planners. The context of the system is vital to the workability of what is being proposed. For example, a complex mainframe system dependent upon a sophisticated support system would be contextually inappropriate in a remote location with few trained staff.

Data Often unstructured, unverified, unvalidated material thought to be the basic foundation for information products, e.g. sales data is complied in information on company turnover.

Debugging The procedure whereby a new item of software is analysed to make sure that it has no built in faults or 'bugs'.

Decision support system A computer-based system that provides users with the capacity to carry out 'what if' analysis (e.g. 'If I increase inflation by 4 per cent and decrease output by 7 per cent, what does this do to my balance of trade?). Decision support systems or DSS are usually based on spreadsheets.

Decomposition The process of working out functions to sub-functions and if necessary sub-sub-functions.

Dialogue systems The screen menus that a computer uses to instruct users and explain procedures.

DOS (disk operating system) The software that controls the user interface to the computer and the working of the computer. Used in Chapter 7 to indicate the wide range of microcomputers that use the industry standard operating system MS-DOS.

Drive alignment Refers to the ability of the disk drive to read the floppy disk. If the drive is out of alignment, then there will be problems with both reading and writing data.

Eclectic methodology An approach to problem solving with systems analysis and systems design methodologies where a number of different approaches are brought together, linking the best or most appropriate parts of each, in order to arrive at a new, mutant approach. Multiview might be said to be an eclectic methodological framework.

Entities The description of anything about which we wish to keep records.

Error messages Often incomprehensible message (e.g. BDOS ERR ON B:) sent to the computer screen to inform the user that something is amiss!

Events Triggers that will spark off functions in an organization.

Expert imposition An attitude summed up in the anonymous quote 'I am the computer expert, please let me get on with sorting out your problem without interruptions!'

Fragmented files Files that have been separated into many small pieces by software process.

Functions The activity that is associated with entities in information modelling.

Hackers Any individual using a computer-based information system without authorization.

Hard Term used to denote a technocratic approach to problem solving.

Hard copy Usually refers to the printed output from a computer system. Hard copy comes in the form of reports, draft documents, letters, maps, etc.

Hard disk A fixed disk drive that is usually located in the computer and is not designed to be removed. Hard disks are usually large, ranging from ten to several hundred megabytes of storage.

Hardware The computer unit itself and its related components–this usually works out as the visual display unit or monitor, the main processor box, a keyboard, printers, modems, fax machines, and optical scanners.

Human activity system A concept created by Professor Peter Checkland used in the soft systems approach (SSM) to systems analysis and systems design. The human activity, concept includes initially understanding the problem context, discovering central themes with the root definition, and setting out proposed improvements.

Human–computer interface That which makes the connection and interaction of the human side of an information system and the technology as easy as possible.

Information A product derived from data assumed to provide knowledge to facilitate decision making, e.g. company turnover information, derived from sales data, provides the basis for making knowledgeable decisions.

Information modelling A hard analysis and design technique. The technique is concerned with setting out the major entities of a system, their functions and attributes, and events that will trigger the functions into activity.

Information systems Any system that provides the user with the stimulus to action.

Install software All software needs to be set up or 'installed' on a computer before it can be used. Installing usually is the method for making sure that the software makes best use of the computer processor, screen type and printer.

Isolated computing The first epoch of the computer age. Firmly related to mainframe computers.

Knowledge The results of good information. Knowledge is the basis of good decision making, e.g. knowledgeable decisions arising from company turnover information will affect the overall strategy of the company.

Login/logout The term for getting into and out of a computer system.

Lost clusters A cluster is a group of sectors on a disk allocated to a file. Lost clusters are just that.

Mainframe A large computer, usually occupying an entire room. This type of computer requires a very highly controlled environment and several trained operators under the authority of a computer manager.

Management controls and constraints The third epoch of the computer age, referring to the period where computer functions begin to be harnessed by organization managers.

Management information systems (MIS) MIS are usually based on computers but do not have to be. Usually they are designed around the idea of supplying management at set times or on demand with key information products called performance indicators.

Menu-driven system The term used to describe a computer-based system that works according to users selecting a series of menu options in predesigned sequence in order to arrive at the information products required.

Methodology A term used to refer to an approach or system for planning information systems.

Mice or mouse A small, oblong box with a ball in it (or an infra-red plate) used to move the cursor on the screen by making contact between the ball (or plate) and a surface. Moving the 'mouse' up the desk moves the cursor on the screen upwards, etc. The mouse has one or several buttons on it. By pressing the buttons when the cursor is close to or on top of an item on the screen (e.g., a word-processed file) you open that item.

Microcomputers The most common computers. Generally these are on the desk top, run common software packages, and are versatile enough to go into quite harsh environments. The most common microcomputers are 'IBM compatible'.

Milestones Significant points for the evaluation of achievement. For example, a milestone for the implementation of a new information system might be 'review progress on data incorporation into the computer system at the end of the financial year'.

Mindset Sorry about using this term but it is quite useful. A mindset indicates all the aspects of your state of mind on a given issue at a given time. If we are working on a rich picture, the picture should represent the mindsets of the analyst and the major stakeholders involved in producing the overall diagram.

Multi-user A computer system with a network that allows a number of users to use the same software at the same time.

Multiview An eclectic methodology originally designed by A. T. Wood-Harper.

Network The name for linking together computers so that they can share information. Popular networks include Ethernet-based systems, and UNIX systems.

Partition The term used to describe separating a hard disk into several different logical disks. For example, the hard disk might be called drive C:. By partition we can make several logical drives (C:, D:, E:) on that single actual drive.

Performance indicators A specific measure (qualitative or quantitative) of the degree of success or failure of a given factor.

Piracy The illegal copying of software products.

Problem situation A real-world information situation in which there is a sense of a feeling that things could be better than they are, or some perceived information provision improvement.

Recipient community That group of users who will be the eventual information system managers, operators and clients.

Reductionist A scientific approach to problem solving that is objective and technocratic.

Rich pictures Pictorial/diagrammatic representations of the situation's aspects: hard and soft: structures, processes, relationships, issues and tasks.

Role of the user The fourth epoch of the computer age. At last the term 'user' is adopted as being important. Machines begin to be oriented towards user requirements.

Root definitions Concise verbal definitions expressing the nature of purposeful human activity systems regarded as relevant to exploring the problem situation.

Screen burn If a computer is left on and not used for a considerable time, the image on the screen will eventually become burnt on. Thus, you will be able to see what has been on the screen even when the computer is off.

Social and technical The linking together of human and technology resources to systems or requirements to make the best fit combination for the specific problem context.

Soft Term used to denote a social sciences, empathetic approach to problem solving.

Software Refers to the application (e.g. word-processors, databases) that run on the hardware.

Software–hardware mismatch The wrong software on the wrong hardware.

Stakeholders Any individual with an interest in the existing or proposed information system.

Stand-alone A computer that works independently from other systems, i.e. a computer that is not networked to other systems.

Systemic An approach to problem solving that is based upon the social sciences. Pragmatic and often subjective.

Systems analysis and systems design The process of discovering what an information system should do and setting out a plan for a workable solution.

Technical subsystems The range of computer systems that combine to produce the total information system.

User–machine interface The latest and fifth epoch of the computer age. The focus is the communication between the user and the machine.

UNIX A computer operating system originally designed by Bell Laboratories. Long thought of as the natural system for minicomputers, it is increasingly being adopted for microcomputers.

Viruses Software, usually designed for fun, that will make a computer break down in a minor or major way. Viruses are usually brought into computers on pirated software. They can lead to total system collapses.

WAN (wide area network) A system that ties local computer power up to international computer systems such as databanks.

Web An approach in systems analysis and systems design that attempts to see all aspects of an organization (management systems, financial systems, information systems) as an integrated whole. Web is generally seen as being opposite to discrete entity analysis, which sees all systems within an organization as separate.

Word-processing The most commonly used form of modern computing. Much more than clever typing, this software application provides users with the ability to input ideas, develop themes, write, work, and rework text and usually produce top-quality printed output as well as draft.

Workspace The amount of store or memory or time or all three that is allotted to a user.

Worldview The overall set of assumptions or highest point of abstraction for any one individual or organization (also known as *Weltanschauung*).

INDEX